To Dad
Happy Christmas 1993

# JAGUAR
## PERFORMANCE AND PRIDE

BY PETE LYONS
AND THE AUTO EDITORS OF CONSUMER GUIDE®

Publications International, Ltd.

## Owners

*Special thanks to the owners of the cars featured in this boo r their enthusiastic cooperation. They are listed below, along the page number(s) on which their cars appear:*

**George E. Alleger**—187 **Jack Bart**—121 **Robert Budlow**—171, 179 **J. Kim Callahan**—142, 143 **Edward Cline**—158, 159, 161 **Dort Crabb**—10, 249 **Sherwood H. Davis**—50, 51 **Nigel Dawes**—146, 252 **Richard Foster**—18, 19, 55, 133, 138, 139, 140, 141, 156, 157, 159, 160, 164, 166, 167, 176, 178, 248 **Cliff Gentle**—8-9, 33 **Julie Green**—126, 127, 248 **Ronald and Sonya Halbauer**—9, 26-27, 52, 53 **Torben Stiig Hansen**—144, 150, 151, 182, 183, 185 **Martin Hilton**—107, 112, 113 **Scott Holdlmair**—96, 97, 98, 99, 104, 105 **Holly Hollenbeck**—47, 57 **Richard P. Howe**—138 **Donald Kamm**—66, 67 **Larry and Susanne Knudsen**—70, 73, 82, 134, 135, 136, 137, 141, 142, 246, 247 **Kay L. Kratz**—191 **Commander Marr**—172 **Dennis Murphy**—66, 67 **Gerald and Katherine Nell**—7, 46, 118, 119 **Carol Nesladek**—162, 163, 167, 168-169 **Gerald Patnode**—124 **Frank Piggot**—64, 74-75, 80-81, 152, 153 **Kenneth Pollock**—39 **Ronald R. Schneider**—6, 25, 58-59, 69, 70-71, 177 **Bruce Sears**—154, 155 **Walter J. Smith**—174, 175 **Judy and Len Smyth**—148, 149 **B.S. "Bib" Stillwell**—116, 117, 252 **Phil Taxman**—28, 29, 172-173, 250 **Lorraine and Bob Tracey**—6, 248 **Doug Zwack**—180, 181

Cover Credit: **Richard Foster, Ronald and Sonya Halbauer, Larry and Susanne Knudsen**

## Photography

*The editors gratefully acknowledge the cooperation of the following people who have supplied photography to help make this book possible. They are listed below, along with the page number(s) of their photos:*

Special thanks to: **Colin M. Cook, Jaguar Cars Limited, gland and Michael L. Cook, Jaguar Cars Inc., USA**

**sport**—216, 252 **Scott Brandt**—184 **Henry ark**—162 **Gary L. Cook**—126, 127 **Bill Coby**—2-173, 250 **Mirco Decet**—8, 14, 15, 17, 25, 30, 39, 83, 94, 95, 98, 106, 107, 110, 111, 112, 113, 115, 120, 121, 123, 124, 125, 128, 129, 130, 131, 147, 196, 197, 198, 199, 200, 201, 202, 210, 212, 213, 243 **Patrick J. Eggs, TWR JaguarSport**—204 **Steen Fleron**—8-9, 33, 122, 123, 144, 150, 151, 182, 183, 185 **FotoFanatics**—11, 218, 224, 225, 227, 228, 229, 230, 231 **Mitch Frumkin**—251 **Thomas Glatch**—6, 7, 46, 58-59, 69, 70-71, 118, 119, 122, 171, 177, 179, 191 **Sam Griffith**—162, 163, 167, 168-169 **Hidden Image**—235, 244 **Ron Hussey**—10, 249 **Bud Juneau**—66, 67 **Milton Gene Kieft**—9, 26-27, 50, 51, 52, 53, 70, 73, 82, 134, 135, 136, 137, 141, 142, 246, 247 **LAT Photographic**—218, 226-227 **Dan Lyons**—6, 121, 124, 165, 248 **Pete Lyons**—44 **Vince Manocchi**—47, 57, 180, 181, 251 **Doug Mitchel**—39, 64, 66, 67, 68, 72, 74-75, 79, 80-81, 96, 97, 98, 99, 104, 105, 116, 117, 138, 152, 153, 154, 155, 207 **PininFarina S.P.A.**—234 **Chris Poole**—192 **Quadrant Picture Library**—208, 218, 220 **Richard Spiegelman**—50, 174, 175 **Denis L. Tanney**—243 **David Temple**—142, 143, 148, 149 **Andrew J.A. White**—170-171 **Nicky Wright**—18, 19, 55, 84-85, 92, 93, 100, 101, 102, 104, 133, 138, 139, 140, 141, 146, 156, 157, 159, 160, 161, 164, 166, 167, 172, 176, 178, 187, 247, 248, 252

Cover Credits: **Mike Cook, Jaguar Public Relations, Milton Gene Kieft, Doug Mitchel, Nicky Wright**

Louis Weber, C.E.O.
Publications International, Ltd.
7373 North Cicero Avenue
Lincolnwood, Illinois 60646



**Printed and bound in Yugoslavia.**

8 7 6 5 4 3 2 1

**ISBN: 0-88176-983-5**

Library of Congress Catalog Card Number: 91-61371

# CONTENTS

# INTRODUCTION: JAGUAR, BURNING BRIGHT

I wonder if my own experience parallels that of others: I liked Jaguars before I liked cars.

It was a long time ago, years before I learned to drive, when cars were boring boxes that took too long to get somewhere. But my father loved the things, and one day, on one of those tedious expeditions in the family De Soto to the sports-car races at Bridgehampton, way out on New York's Long Island, we broke the trip at a lonely hilltop hostelry called the Scotch Mist Inn. There in the parking lot, crouched like wild animals, was a small pack of the first interesting cars I had ever seen.

They're Jaguars, I was told. XK 120s. They're sports cars, and they come from England. Don't be too long, we have to eat.

I walked around them, and around them again. What exotic, beautiful, compelling shapes. What an aroma they secreted of excitement, daring, wealth, romance, adventure. What wonderful automobiles!

From then on, though an actual passion for cars in

*Coming or going, Jaguars always made a statement. Their lines bespoke a sensuality, certainly, but the clear message was one of prowess as much as passion. The XK 120* (below and left)*; the E-type Series II* (bottom, both photos).

general had yet to be ignited, the Jaguar car was always smoldering somewhere in the back of my head. And even when I finally had come to an adolescent incandescence over sports cars, and had discerned many that were more desirable than Jaguars in various ways, still there was something about the kind of sports motoring I saw in a Jaguar that fed a special flame.

In a sense, I feel I grew up with Jaguars. I really think I remember every one I've ever seen. An XK 140 coming out of a country club one sunny suburban afternoon, every chromed spoke of its wire wheels flashing diamond-fire. Another 140 with a supercharger—the first one of *those* I'd ever seen—nestling alongside the gleaming, polished cam covers in the crocodile jaws of its "bonnet." A white XK 150 roadster in which racing driver Walt Hansgen hurled me around Lime Rock Park in Connecticut, my first-ever laps at speed around a track. I remember his calm mastery of the big black steering wheel as an astonishing storm of *g* forces battered my body. And another white 150 roadster, this one decked out in a dashing red racing stripe, where I acted as navigator in my first road rally through the forests south of San Francisco. I found I enjoyed rallying all right, but what moved me to rapture was that car's fabulously savage exhaust note.

My first E-type. It sprawled under spotlights at the New York show like a great lazy lioness. I was helpless prey. I positively ached to be possessed by that automobile. A few months later, I encountered my first one on

*Sometimes a Jaguar was visually and mechanically "dated" compared to the best of its contemporaries. It was as if Coventry valued the themes of the ages over the fashions of the day. This is a 1959 XK 150.*

*A surprising variety of Jaguars—sedans and roadsters alike—tested their mettle on the racetrack. A V-12 carried this XJC coupe into battle.*

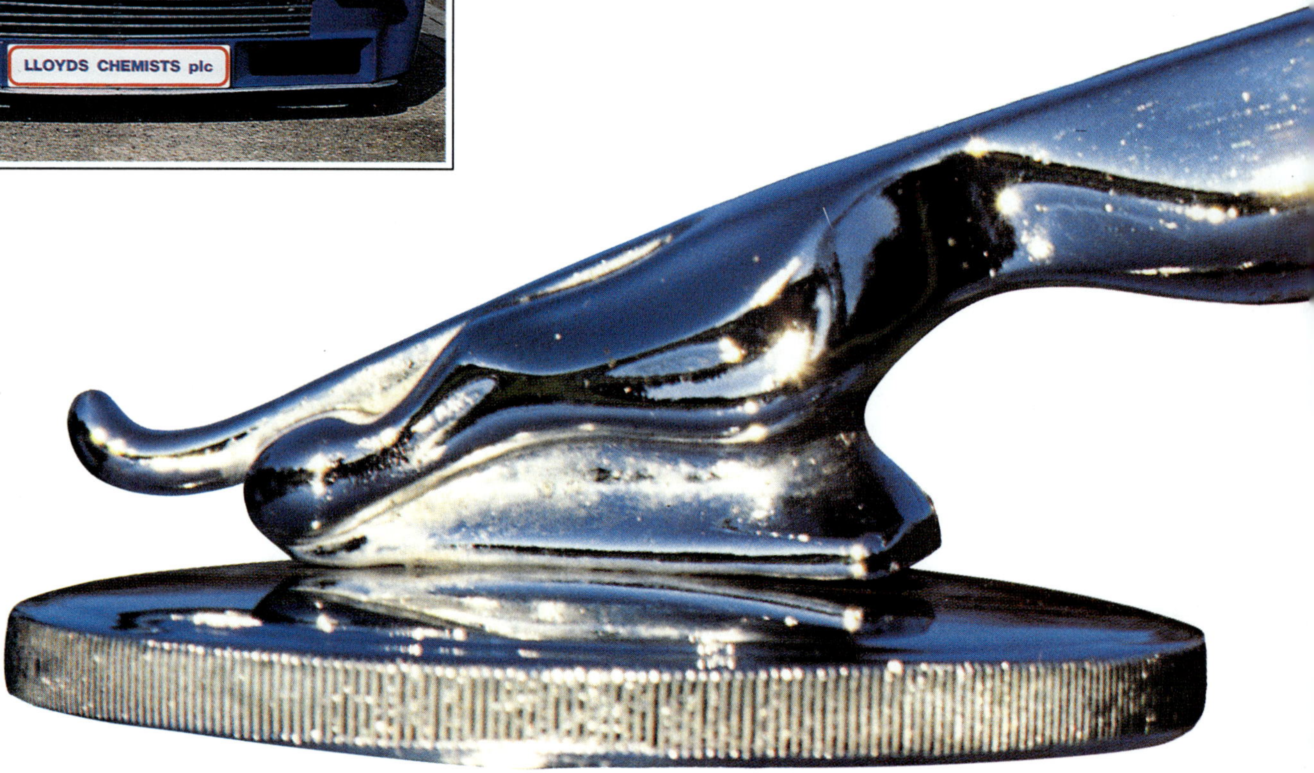

the road. It was an evening in London, and I was just banking my newly acquired Norton motorcycle into a sweeping Hyde Park Corner when a dark bomb-shaped thing shot by under my elbow. I sprang to the challenge and tried to keep up, but I lost. Badly. What a car.

What is it about Jaguars? As automobiles they have always fallen some way short of seamless mechanical perfection, yet they retain some transcendent, mysterious appeal. And it appears to be a universal appeal; Jaguars will turn the heads of people who care nothing about cars just as readily as they stop true believers in their tracks. This is not unique in the automotive world, but it is rare, and Jaguar pulls it off with supreme panache.

To the Brits, I think, the Jaguar is more than a motorcar. Much of the Jaguar driving I myself have done (finally!) has been across the landscape of its birth, and I've seen repeatedly that the marque draws a local reaction both strong and unique. Other drivers will smile, and give a thumbs-up. Pedestrians will cross a road to start a conversation, and ask to see under the bonnet. I've actually had an English policeman guarding a car-park against lesser vehicles move a barrier, unasked, to let my borrowed Jaguar in.

For some people of that island nation, I've come to suspect that their beloved Jaguar has a kinship with a Battle of Britain fighter plane. A Supermarine Spitfire, perhaps, another great hunting machine speeding out across the Channel in stalwart challenge of the foreign foe. The Jaguar has also brought combat victories home to England, not only from Le Mans and so many other international races, but from sales battles around the world.

*Lyons' Swallow works put bodies on the most mundane mechanicals; this '36 SS1 Tourer took 23 seconds to reach 50 mph. But there was about them an appealing roguishness.*

When the British see a Jaguar, they see a British winner.

They see beauty, too, and that's really the combination we all find so compelling about Jaguars: That feline blend of grace and balance with muscle and sinew, of civility with savagery. And Jaguar had the wit to stir the mixture not only into its sports models but into its sedans and 2+2s so every Jaguar driver could taste it: As with the XK 120, so with the stealthy Mark VII sedan; as with the XK 150, so with the graceful Mark II saloon; as with the E-type, so with the lissome XJ-6 four-door. To be a driver of any of these great Jaguars is more than to own a certain make of car.

The Jaguar of today still largely represents the life's work of one gifted man, an amazing combination of salesman and artist. William Lyons—to whom your author is no relation, alas—was not a mechanic and certainly no engineer, yet he brought about the creation of machinery that frequently pierced known engineering boundaries. He built his cars with eyes fixed firmly on style and cost, but time after time saw them prove their performance and reliability in the world's most grueling speed competitions. In business, Lyons had to face many hard decisions, yet he managed to hold the loyalty, often fierce, of his staff—and the love, literally, of millions of his customers. He even earned the ultimate appreciation of his nation, which knighted him in 1955. In design, Sir William worked directly, standing by as craftsmen shaped metal to his command; then he'd have the prototypes brought to his estate, where he and his wife, the Lady Greta, could evaluate them in properly elegant, Jaguarish surroundings. (Visitors to the Lyons home once found Sir William industriously arranging and rearranging baggage on his hallway floor; he was, he explained, working out the ideal dimensions and volume of the "boot" of his next "saloon.")

William Lyons was at once Jaguar's strength and its weakness. No one else could have accomplished what he did, not only in terms of car design but in commercial daring, in holding the firm together in times of trial. At the same time, it must be admitted that Jaguars reflected his own range of tastes and interests, sometimes to the exclusion of qualities of concern to others. And he guided the company itself with such a strong, solo hand that when he was gone, the reins fell slack for many years.

Along with others of the great marques, values of many older Jaguars have risen sharply of late. That's as deserved in this case as any. Yet it would be wrong to lose sight of the darker side of Jaguar ownership, the oil-stained driveway side. Jaguars of the period we now regard as the "classic era" were built down to a price, and there is some truth to the old saying that the buyer was expected to finish building the car. The more complex models demanded expensive maintenance, but because of their aggressive pricing they often came into the hands of people who could ill-afford to keep them up correctly. Many owners did manage, and have joyously owned Jaguar after Jaguar. But there are those who once succumbed to a cat from Coventry and say it has forever soured them on every English-made car.

It would be equally wrong, however, to overlook the very real achievements of Jaguar in competition, especially at Le Mans. Five outright victories in the classic period, plus two more (so far) in modern times in what is still the world's premier endurance contest, prove a raw-

*The SS1 of '33* (above and right), *the spry Austin 7 and the fierce E-type* (top) *all bore Lyons' stamp*. Opposite, top*: XJS Cabrio and SS 100.* Opposite, bottom*: Daimler Sovereigns.* Opposite, right*: Victory at Le Mans!*

boned strength in both the machines and the men who made them.

Such dichotomies well characterize Jaguar's history. To trace the whole long, nearly seventy-year tale is to notice curious shifts and changes and inconsistencies. It's hard to single out one unvarying trait and say "Jaguars are . . ." The SS1, Lyons' first complete car, was a rather odd vehicle, not very promising either in appearance or performance. Yet it shortly gave birth to the very impressive SS 100, a genuine 100-mph sports car in a period when such a claim was hard to prove. The lithe XK 120, which captivated so many, as it did me, was allowed over the years to become the "fat cat" XK 150. But then, having been kept on long past its technological prime, having finally become distinctly archaic, the 150 gave way overnight to the spectacularly advanced E-type.

The fact is, Jaguars are as much about sensuality as science. At their best, they blend serious road performance with style and romance and passion; how odd that they issue from a nation so widely perceived as reserved, bland and cold. By their Jaguars, the British stand exposed as enthusiasts almost as ardent as the Italians.

No one has ever purchased a Jaguar to fulfill a simple requirement for transportation. The choice is made on other grounds: fashion, grace, elegance, pride, prestige, speed, excitement, joy. Jaguars are not about getting somewhere, but going. Going nowhere, if you're really lucky, merely driving at a canter through a brisk country day, top down, with that savage engine singing.

Jaguar is an automaker that exists not because anyone needs automobiles, but because they love them. Not an ignoble flame to keep alight.

*Pete Lyons*
*Big Bear, California*
*March 1991*

CHAPTER I

# A MAN AND A DREAM

There is such a marvelous aura of inevitability about the Jaguar that it is hard to realize this widely beloved marque came about by chance.

As with most of the great automobiles, Jaguars primarily reflected the vision, taste, ability, perseverance, and strength of character of one man. Yet William Lyons, co-founder before his twenty-first birthday of the very small motorcycle sidecar factory that would grow to become Jaguar Cars, Ltd., was never really a "car guy" in the engineer-mechanic mold of Karl Benz, Gottlieb Daimler, Charles S. Royce, and so many other pioneer automakers. Lyons made his name among them honestly, but coincidentally. He was by no means disinterested in automobiles, but as a young man he was more interested in motorcycles, and even more in launching a business. Any business. Although he had early experience in the motor-vehicle trade, it was pure happenstance that led him into sidecar manufacture. Though he eventually would be knighted for his accomplishments, he could have had no precognition of the ultimate magnitude of his enterprise.

This is not to belittle his genius. The son of a professional musician, he had no formal training in either engineering or styling. Rather, the cars he came to build were shaped in both areas by his own intuition, and this proved to be in uncanny harmony with the wants of his public. The classic Jaguars all bore Lyons' personal stamp. It was their strength and, it must be said, their weakness as well.

It is surely germane to the Jaguar story that William Lyons was born in a pleasant place, at a happy time, to good people. The scene was Blackpool, a popular resort and convention center on the northwest coast of England. It was, and is, a town of generous beaches, a seaside promenade seven miles long, and numerous commercial entertainments, all dominated by a quasi-replica of the Eiffel Tower. To this minor cultural mecca toward the close of the nineteenth century came an enterprising Irishman named William Lyons, then a member of a touring orchestra. He decided he liked Blackpool; even more, he decided he liked a Blackpool lass named Minnie Barcroft. They married, and William continued his performing and writing of music, but he added depth to his affairs by opening a business he called Lyons's Music and Pianoforte Warehouse. The first of two children was Carol; then on September 4, 1901, came William Junior.

Billy was a goodly lad, athletic in build, less apt at academics than sports. It seems he especially enjoyed running, at which he won a school prize. He also contracted a passion for motorcycles. By the time he left school, at the age of 17, he was buying and selling the exciting things, and racing them a little as well. On such bikes as Nortons, Indians, and a Harley-Davidson model known as the Daytona Special, which he called his favorite, he competed in speed trials on the local beaches and in nearby hillclimbs. He won at least one such event, at Waddington Fells in the nearby Forest of Bowland. Photos taken at this period show Bill Lyons as a long, lean, well-groomed young man with a notable nose, eyes brimming with hawkish intelligence, and rather sensitive lips fixed in firmly self-confident determination.

That last trait came out when his father tried to settle him into a respectable career near the close of World War I. Junior, who showed no interest at all in the family piano business, had applied to be taken on as an apprentice shipbuilder. William Senior thought he saw a better opportunity, and fixed him up with Crossley Motors, a respected automaker of the time in Manchester, a regional center some 50 miles inland—a not insignificant distance in 1918.

"I found it a rather depressing, plodding existence," the future Jaguar man told marque historian Andrew Whyte many years later. "I was 17, and old

*Jaguar's founder had a young-man's passion for motorbikes and little appetite for schooling in design or engineering. But few possessed as pure a sense of automotive style or value. Sir William Lyons, 1901-1985.*

FR.3030

enough to realize that the war was affecting everyone's lives. I couldn't help feeling a need to reach some personal goals quickly, though. Almost as soon as I started my apprenticeship in Manchester, and going to night school, I realized I wasn't going to stick it for long. Then came the armistice, and the possibilities seemed to broaden. My father wasn't at all pleased when I said I'd had enough of industry, but he let me come home and I even worked in the music shop for a while."

The young Lyons now spent some time pacing impatiently. He was desperate to get something of his own going, but what? He put in his time at the piano repair shop, and talked of opening his own business in gramophones, the fashionable home entertainment of the day. Instead, he took his motorcycle trading skills into the four-wheeler world and joined a local dealership selling Sunbeam, Rover, and Morris cars. In November of 1919, only just 18, he was sent by his employer to far-off London—200 miles!—to help work the first postwar Motor Show.

This was the period when Bill Lyons bought his first automobile. It was used, a prewar Buckingham, a small vehicle that had gained a sporting reputation. But this one proved unreliable, and he soon disposed of it in favor of another motorcycle.

And then William Walmsley's family moved in across the street from the Lyons'. Nine years older than Bill Lyons, Bill Walmsley also still lived with his father, Thomas, a Manchester-area man who had become substantial in various aspects of the coal business. The younger Walmsley didn't much care for coal, but he did find himself fascinated by the carts and railroad cars constructed in his father's shops. It turned out he had quite a flair as an inventor and craftsman.

He also liked motorcycles. Returning from service in the Great War, he brought along a military-surplus Triumph. He was by no means the only demobilized soldier to do so, and soon he had set up a small business to refurbish similar machines. He worked out of a shed in the family backyard, or garden, and it was here during 1920 that he built up a sidecar of his own design for his own bike.

*Lyons' desire to establish himself in business dovetailed nicely with the small sidecar enterprise of a Blackpool neighbor, William Walmsley. Walmsley had designed a nice Zeppelin-like sidecar, but it took Lyons to fully mine its commercial potential.*

Motorcycle-sidecar "combinations" have never appealed to a people more strongly than to the British, and the post-World War I period was one of the contraption's great heydays. Part of the reason stemmed from the nation's taxation policies, which made motorcycles especially economical forms of personal conveyance. But the weather typical of the British Isles, especially in winter, lent a particular attractiveness to a third wheel, not to mention an enclosed cabin. Perhaps the inherently tricky technique involved in sidecar driving, where left and right turns are negotiated differently, had something to do with it, too: The Brits are a sporting race.

Walmsley is remembered as a straightforward, low-key fellow with a hearty, if often alarmingly blunt sense of humor and a dashing appeal to the local girls. It is generally supposed the latter interest led to his interest in sidecars. Most combination bodies of the day were strictly utilitarian, but he designed his as a sports sidecar. The body was fashioned of eight segments of sheet aluminum, each cut in a petal shape so that when assembled on their framework of ash timber they formed a pointed-nose fuselage reminiscent of a dirigible airship. The bodywork, its aircraft-like open cockpit upholstered and trimmed by his three sisters, was mounted on a proprietary sidecar chassis, the single wheel of which was streamlined with a polished aluminum disc. When he was done, Walmsley stepped back to survey his gleaming bullet, and was so pleased he gave it a risqué nickname: "Ot-as-ell."

It was irresistible, of course, and doubtless within minutes of its first appearance he had orders for duplicates. On a very low-key scale, Walmsley converted his motorcycle repair shop in his parents' backyard shed to a sidecar factory, employing various family members, including his wife, Emily, to help out. As a gesture to gentility, he called his production models "Swallows."

In June of 1921, having cashed out of his coal business, Thomas Walmsley retired to an upscale neighborhood in Blackpool. His son tagged along, and the two-car garage behind the new home was soon humming. A 19-year-old motorcycle enthusiast with an eye for good design who lived across the street could not

*Lyons was an eager 20-year-old, Walmsley a rather unambitious 30, when, with financial help from their fathers, they launched the Swallow Sidecar Company in 1921 The future Jaguar man plunged ahead, pursuing sales and growth with a zeal Walmsley didn't share.*

*By the mid twenties, the Swallow works was humming, turning out sidecars of various styles. Production efficiencies allowed periodic price reductions and helped it weather competition from England's new generation of small, inexpensive automobiles.*

help but notice what was going on. Lyons soon made Walmsley's acquaintance and placed an order for a Swallow of his own.

Except for one thing, what happened next had that touch of inevitability that would come to characterize so much of Jaguar history. The situation was well-nigh perfect: There was young Bill Lyons seeking a business opportunity to sink his teeth into, understanding both motorcycles and marketing, and seeing a highly saleable product in neighbor Walmsley's sleek Swallow sidecar. But he also observed that Walmsley, who had demonstrated his design and craftsmanship talents, was not doing the job on the advertising side. Most importantly, both successful fathers were willing, not to say anxious, to back their unfocused sons in a promising venture. Meeting at a local bank, the seniors signed a loan guarantee of 1000 Pounds Sterling to launch the fledgling Swallow Sidecar Company.

The bristle in the paint? The new partners really didn't get along very well. Lyons and Walmsley may have shared the same Christian names, but that, and their mutual enthusiasm for bikes, was about the extent of their compatibility. Bill Walmsley, a shambling, shapeless fellow of irreverent wit and rather limited outlook, never did accept the urgency of the younger man's ambitions. Evidently, his family's economic status engendered in him no desperation about income. Then, too, his health was not robust: Besides having injured a leg during his service in the army cavalry, he suffered lifelong from pernicious anemia. Creative, but not driven, young Walmsley would have been perfectly content continuing as he had been, handmaking a handful of sidecars in his father's garden.

But young Lyons' drifting years were over. He gave notice at the car dealership and plunged into "his" new business with messianic zeal, drumming up sidecar orders, signing dealers, placing ads and printing brochures, looking for a building to rent so production could be stepped up, hiring workers, handling the office

*Lyons' promotional talent was in evidence early on. Swallow sidecars were entered in races and happy customers' letters were used in ads. But all was not hype; design and workmanship were of a high order.*

paperwork, organizing a Swallow presence at the 1922 London show, putting in very long days, and planning, planning, planning.

Although conventional wisdom held that the great motorcycle sidecar boom had ended, victim of England's new generation of small, cheap automobiles introduced in the 1920s, Swallow prospered. Most of that was no doubt due to William Lyons' tireless enterprise, but some credit has to be given to William Walmsley's knack for design and production. Their company introduced a variety of successive models, some developing the attractive Zeppelin-like styling theme, others exploring more conventional configurations. Swallow sidecars became prized for their blend of outstanding looks, satisfying luxury, decent quality and good value. The young firm even took care to develop something of a sporting image by placing its lighter models in various competitions. On the other side of the performance coin, a number of Swallows were sold to a large police department. Testimonials came in from satisfied customers, who were liberally quoted in promotional literature.

Image was one thing, but Lyons knew he had to offer value as well, and a serious-minded approach to production efficiencies enabled periodic reductions in prices. Sales kept soaring, the bank kept advancing money, and the Swallow Sidecar Co. kept having to expand into a succession of makeshift premises in Blackpool.

Staffing grew apace. The original half-dozen employees doubled, then doubled again. From memoirs of old hands, Swallow was a happy place. The cramped quarters, primitive working conditions, and heavy workload, natural consequences of William Lyons' tight-ship approach to funding, were offset by a heady sense of backing a winner. There seems to have been a strong team spirit, even a warm family atmosphere among those early Swallow people, and a remarkable number would prove faithful to Billy Lyons for decades to come.

CHAPTER II

# A Marque is Born

Just which William may have been more responsible for the next step on the road to Jaguar is an interesting question. Both men remained keen motorcyclists, keen enough to run a matched pair of Brough (pronounced "Bruff") Superiors, the class British bike of the period. In fact, they acquired two pairs: SS80s first, then when Brough brought out the more powerful SS100—tuck that model designation away for future reference—they went back for a pair of those. But, as will happen to young motorcyclists, reasons began to mount up for owning an automobile; even George Brough, in his turn, would later become an ardent Jaguar owner. Walmsley, already married, now had a small son, Bobby. Lyons began to get serious about a young lady named Greta Brown. She seems to have cheerfully put up with his courting her by motorcycle, and accepted his invitation for a holiday trip to Scotland in his spiffy Swallow sidecar. They were accompanied by a car, a Studebaker, full of her family, of course—it was 1923, after all. But when William and Greta were married the next year, and went back to Scotland for their honeymoon, they traveled on four wheels, his own father's Talbot.

Not long afterwards, he realized he had to buy an automobile of his own, and that second car owned by

*William Lyons' first automobiles used bodies of his own design atop the chassis and running gear of the tiny Austin Seven. Below is his Austin Swallow Saloon of 1928.*

William Lyons may have been seminal. It was an Austin Seven, one of the new British babies that were transforming the very concept of automobile ownership because they were so cheap. Cheap, yes, but the parsimonious Lyons bought a two-year-old one. He doesn't seem to have been especially fond of it, and presently he moved on to larger vehicles, such as Alvis and Vauxhall. But the little Austin stayed in his mind.

Meanwhile, entirely on his own, Walmsley acquired the burned-out chassis of another car, an old Austro-Daimler, and got a couple of his company's craftsmen to help him construct a new body for it. Although little public notice was taken of this hobby-project at the time, it can be regarded as the first Swallow automobile, and it is difficult to shake the notion that the inventive William Walmsley in his quiet, straightforward way already had something bigger in mind. Perhaps he invited his wealthy father out for a comradely pint of bitter beer. Or maybe Lyons did the talking. At any rate, it was in 1926 that Thomas Walmsley purchased a large, two-story commercial building on Blackpool's Cocker Street and made it available to his son and his partner. Not only was it big enough to consolidate Swallow Sidecar's three separate, scattered facilities, it was large enough to house an additional venture: an automotive body shop. Thus by that November, not yet five years after starting up their sidecar business, Lyons and Walmsley altered its identity to the Swallow Sidecar and Coach Building Company.

At first, the automotive operation advertised only the repainting of cars, plus the refurbishing of their interiors and folding tops. But one day that winter, someone brought by a sorry-looking race car, a Talbot-Darracq, that had been rolled during a meet on nearby Southport Sands. Could Swallow do something with it, please? Swallow could, because among a number of body specialists it had recently hired from the big automaking plants in the English Midlands, around Birmingham, was Cyril Holland. A native genius at design as well as woodworking and metalsmithing, Holland started sketching some body shapes.

According to the stories told in later years, Lyons came by his new man's drawing board one day to see what was going on, was captivated by what he saw, and on the spur of the moment told Holland to re-dimension his drawings to suit an Austin Seven. Then he went to his phone to cajole a regular Austin dealer into breaking corporate rules and selling him a bare chassis.

Certainly the idea of replacing a stock sedan body with something more appealing did not originate with Swallow. In 1920s Britain it was called "packaging." Throughout the industry, chassis and coachwork were still thought of as separate items, and specialist coachbuilding had long been established amongst the gentry. In fact, it was not uncommon for really wealthy types to buy two bodies for swapping as needed on a single chassis: an open style for summer, a snug closed type for colder weather.

By the late twenties, however, there were firms working with cars and customers a bit lower on the social scale, and even a couple of Swallow's sidecar rivals had started re-bodying small chassis. Also, Cecil Kimber's sporty MGs, then starting to make their name in racing, were based on Morris components (MG being short for the Morris Garages company). No recluse, Lyons would certainly have been aware of all these trends. In fact, a friend of his once brought by an Austin Seven wearing a sporty new fabric body by Gordon England.

But such projects were usually done on a small scale, even as one-offs. However they came about it, Mssrs. Lyons' and Walmsley's plan reached farther than anyone else's. Right from the outset they set up for mass production, and successive models appear to have been an original part of the plan.

The first Austin Seven-Swallow made its debut in May 1927. An attractive little thing even today, it must have been a real head-turner back then. Austin's normal Seven, Britain's popular equivalent of the Ford Model T, was not exactly ugly, but it was deeply utilitarian. Swal-

*Jaunty styling, low prices, and paint schemes that stood out from the grays and browns of the day diverted attention from the humble specifications.*

*The Austin Seven-Swallow of 1927* (right) *was a pert two-seater with a tiny 748-cc, side-valve, four* (above). *It sold for just £175, and sparked Lyons' automaking enterprise.*

low's replacement body was pervaded by a note of cheery frivolity, being a pert two-seater with neatly rounded lines, harmonious proportions, and a bright cream-and-crimson color scheme. The interior was finished off in good materials, too, and overall the Austin-Swallow gave the impression of being a quality car. It was not yet a sports car, though, because it retained the Seven's primitive mechanical components almost without alteration, including a tiny 748-cc, side-valve, four-cylinder engine. But it did offer a breath of springtime, and at a base price, £175 Sterling, that seemed breathtakingly affordable.

Even Lyons was surprised by what he had wrought. First, he went along to the Blackpool dealership that had employed him briefly years before. He left with an agreement to handle his new car. Then, setting off on a sales tour with an Austin-Swallow, he stopped by a dealer in Birmingham. On the spot, he received an order for 50 cars. Thus encouraged, he drove on to London, where the prestigious Henlys floored him by signing up for 500. How exciting must have been his long trip home as he planned the moves necessary to meet the gigantic task of filling these orders. As one of those moves would be to yet another factory, he could no doubt hear his partner's reaction, which had to be something like, "You're mad, Lyons. Stark, staring mad!"

Not mad, ambitious. Decisive, too, and soon he was making a series of explorations to the industrial Midlands. It was apparent that Blackpool was just too small and remote a place to support an enterprise of the size Lyons was creating. In Coventry, about 150 miles away on the far side of Birmingham, in a district called Foleshill, he finally found a factory building he could afford. A long-derelict plant originally built to load explosives into artillery shells, it was in dismal condition, and located at the end of a dirt road that was usually a bog. But to William Lyons it looked like home, and he was able to

*An important symbol of Swallow's expansion beyond its sidecar-maker roots was its first appearance at an automobile show, the Olympia exhibition of 1929* (below). *The Lyons-Walmsley concern was still very much a coachbuilding one, however; the cars pictured in the photo use—from left—Swift, Fiat, and Standard engines and chassis.*

persuade the majority of his nearly 50 employees—and Walmsley—to pull up stakes with him. The move from Cocker Street to what soon became known locally as "Swallow Road," Coventry, was complete by November 7, 1928.

Scarcely two years into the automotive game, Swallow was a significant player. Only three months after the Austin came out, it showed a larger car on a 1550-cc Morris Cowley chassis, and in 1928 added to the original Austin two-seater a four-seat sedan, or "saloon." Nor was a model lineup of three enough. In 1929, after a couple of tests on Clyno and Alvis chassis, Swallow launched a still-larger sedan based on the well-made but staidly-bodied FIAT 509A from Italy. The same year brought introductions of the Swift-Swallow and the Standard-Swallow, sedans, respectively derived from British cars called the Swift Ten and Standard Big Nine. Next came another open sporty car, this one mounted on a Wolseley Hornet chassis. All this time, sidecar production remained at surprisingly high levels. Swallow was really flying.

Swallow kept on flying mainly with chassis from the Standard Motor Co., Ltd., perhaps because it, too, called Coventry home. The first Standard-Swallow, a 1287-cc four-cylinder car, was joined in 1931 by the longer-chassis Ensign, a Six of 2054 cc. The company's advertising, to which Lyons always paid great personal attention and which was one of the few areas in which he did not begrudge spending money, hyped this model with such adjectives as "magnificent," "marvelous" and "astonishing"; more significantly, it introduced the phrase "The Thousand-Pound Look." In fact, the car was priced at little more than a quarter of that, so with this campaign was cemented in everyone's mind William Lyons' basic marketing philosophy of offering the buyer a lot for a little. That would hold true for as long as he held his company.

How did he do it? Everyone always asked. Billy Lyons wasn't tight, exactly, but he kept his pencil sharp. The tales are legion of ways he found to keep his costs down, almost "at all costs." He wouldn't buy new factory machinery if used did just as well. He wouldn't hire two people if one could handle two jobs. He wouldn't deal with suppliers who couldn't see the promotional value of being associated with him, and set their prices accordingly. He could see no sense in spending money on his cars in locations that wouldn't matter to his customers.

But he seldom cut so fine as to lose anyone's good will, and the bottom line is that he kept his company alive—thriving, in fact—through times that proved fatal to innumerable others.

Because aging Standard had recently received some young blood of the forward-thinking type that ran in Lyons' veins, it now became possible to take Swallow the next step toward becoming a real automaker. Lyons expected that the fad of merely pretty bodies on boring old chassis was going to fade; happily, Standard was willing to supply him with a foundation built to his own specifications. Accordingly, at the 1931 London show that fall, there appeared a most dramatic, not to say melodramatic, new motor car. Rather in the mode of contemporary sporting Mercedes, it featured an engine hood ("bonnet" to the British) so long as to be a caricature, and a disproportionately stubby 2+2 cab nestled between the rear wheels. The whole car, in fact, was rakishly low, a benefit of the exclusive Standard-built chassis. To point up the new look there was a new marque name. Radiator emblems on some of the Standard-Swallows already had featured "SS" as well as the two formal names; this new model bore only the initials. (Strict historians will note that periods were used initially but later dropped.)

Implicit in Lyons' new car, called SS1 (which later gained a smaller running mate, the SS2) was an appeal far more visceral than cerebral. It was not an engineer's car, but a salesman's. According to press coverage at the time, it had grown out of a brainstorming session with personnel at Henlys, Swallow's biggest and most aggressive dealership. Asked to draw up a wish list of characteristics and features, they concentrated on appearance. "Give us," they said according to contemporary accounts, "a car that is low in build, looks slim like a young woman, has a long body, and generally looks rakish." In short, they wanted that £1000 look for, say, £300. Oh, someone added, and how about giving us seats as luxurious as the one our chairman enjoys in his boardroom? Such was the tall order that Lyons tried to fill.

But fill it he did, thanks to a passion for his cars' appearance that would always be as strong—and successful—as the concern he showed for his business. This trait would give William Lyons a reputation as one of this century's major automobile stylists. The renown is justified by the results, but any image of the man poised before an easel, summoning beauty out of thin air with deft com-

*Lyons relocated his Swallow works to Coventry in 1928, taking an industrious crew of nearly 50 with him* (top, both pages). *The Austin Seven-Swallow was followed by a larger convertible built on the Morris Cowley chassis* (left). *The real breakthrough, though, was the stylish SS1 Saloon of 1933* (below).

mands of his pencil, is inaccurate. According to numerous of his lieutenants, the head man could scarcely draw a line. One has said that his myriad fussy little marks made his concept sketches look like cars with fur coats.

In fact, he seldom sketched at all. His way of working was in the metal, life-size, and out-of-doors if at all possible. He'd summon a group of his best craftsmen and describe broadly what he had in mind. They'd set to, cutting and bending and hammering, and he'd keep stopping by to order revisions, paying particular attention to the car's "face," until one day, presto, there was a mockup of the next Lyons masterpiece. It was full-size, so he could walk around it, even sit in it. It was mobile, too, so it could be trucked to his home, where he and Greta could "live with it" for a while in a real-life environment.

When finally unveiled, the styling of the SS1 scored a bulls-eye with press and public. In fact, the car drew such raves that William himself was surprised. Personally, he was more than a little disappointed in the first model. Appendicitis had knocked him out of action at a crucial moment and, in his absence, Walmsley had directed Cyril Holland to finish the first car to suit his own notions of comfort and practicality—which meant the roofline came out a lot higher than Lyons had intended. But the SS1 sold briskly despite this and a number of early quality and reliability problems, plus performance that was really not up to the looks (and a price that, at £310, was slightly over target).

Patently, Lyons had scored an actual automotive triumph. Perhaps he needn't have changed his new flagship much for 1933, but he did. He got Standard to lengthen the wheelbase and alter the rear of the chassis to open up the interior volume. The longer cab section was now in better harmony with that long "bonnet," and the cobby cycle-type front fenders gave way to a more sophisticated sweeping line. The car was substantially revised in other areas too, including the engine bay, where Standard's new 2552-cc, 62-horsepower engine was an option. Presently offered in additional body styles, including a convertible, the SS1 (that was the official designation on all paperwork, although "SS One" appeared on radiator emblems) continued to be a success with fashionable people for several years. Part of the appeal was their car's steady success in the then-popular *Concours d'Elegance* competitions.

Inevitably, some owners tried their SS1s in real competitions, particularly rallies, and the factory itself started entering and doing well in such events as the Alpine Trials. By showing the sinew beneath its cars' sleek skins—and with orders coming in from Europe and America—SS was increasingly regarded as a serious automaker. With this and the fast-growing renown of the SS marque, the company again changed its name in 1934, this time to—what else?—SS Cars, Ltd.

It was also in 1934 that Walmsley built a sports-car prototype on an SS1 chassis. That was fine with Lyons,

*Most of the early cars bore the SS emblem* (opposite, top), *though it wasn't until 1934 that Lyons bought out his partner, William Walmsley, and formally changed his company's name to SS Cars, Ltd. Lyons was moving quickly to establish himself as a wheels-up automaker.*

and the project led the following year to the production of the SS 90. But the older man was spending even more time—and some of the company's material and personnel resources—on an intricate model railroad at his home. Generally, the partners were diverging more and more in their aims and attitudes. That fall there was apparently one dispute too many, and Walmsley allowed himself to be bought out.

The man who invented the Swallow sidecar as a hobby was financially able and temperamentally content to spend the rest of his life on similar pursuits. His main commercial activities involved the small house trailers the English call caravans. For these he developed several interesting innovations, including roofs that would retract for better road aerodynamics. In his low-key way, he was also prominent in promotional activities for a series of trailer companies. His personal life turned sad, for his first marriage, to Emily, ended in divorce, and his only son, Bobby, was killed in a fighter-plane crash in World War II. He remarried, but in later years his health began failing. Walmsley eventually retired back to Blackpool, where he died in 1961 at 68. Lyons sent two deputies to the funeral, but did not go himself.

In January of 1935, the company now solely his, Lyons went public with it, offering stock on the London exchange. The offering netted a healthy £85,000. Two months later he went public with the SS 90. Built on a shortened SS1 chassis, it was a stark, open sports car in the classic British idiom, SS Cars' first attempt at a true performance vehicle. Obviously, style was no longer enough. Substance would have to start figuring as well—and not only because of the SS 90, for Lyons had in mind

*SS Cars set about developing a more mature and substantial sedan. It put an overhead-valve head on the old Standard engine, widened and stiffened the SS1 chassis, and contracted for its first four-door body. The resultant saloon* (below and left) *debuted in 1935 and with an addition to its SS badge* (opposite, bottom), *was the first car to wear the name Jaguar.*

becoming a complete automaker. One avenue to that end was hiring as an outside consultant the famous cylinder-head expert Harry Weslake, who would supervise the company's adoption of overhead valves. Another was to start a proper in-house engineering department, for which William Lyons hired William Heynes.

This third Bill to figure in the Jaguar story is generally considered the second most important person ever involved in the company's development, surpassing Walmsley's 12-year contribution, because he would continue to give a solid mechanical base to Lyons' aesthetic vision for the next 34 years. Heynes' first idea in life was to be a surgeon or veterinarian, but at 19 he found himself apprenticed to an automaker named Humber. By age 31, when Lyons was referred to him by mutual friends, he had developed credentials in many areas of automotive engineering, including engines, transmissions, and suspensions. Beginning at SS Cars, Ltd., in April of 1935, young chief engineer Heynes immediately plunged into the design and development of an all-new model that had to be ready for the pivotal London auto show that October.

This was to be a true sedan, with four doors and enough length to carry five people in comfort and elegance for extended journeys. In keeping with its intended market, it was given somewhat more restrained, conservative styling very similar to that of contemporary Bentleys, though it still had the recognizable Lyons flair. There were several mechanical improvements, too. But what makes this 1935 vehicle, not in itself especially spectacular, of particular historical interest is that it was the first to carry the name Jaguar.

CHAPTER III

# Sir William and His Great Cats

In coming up with the SS marque, William Lyons may have been trying not only to distance his company from its old "packager" image, but perhaps also from the delicate inoffensiveness implied by the word Swallow. He cannot have forgotten the powerful SS-series Brough Superior motorcycles he'd owned, nor was he likely oblivious of the imposing Mercedes "Super Sports" models of the late 1920s. He may well have thought that the hard sibilance of that doubled-S sound conveyed the feeling of sleek strength he wanted to project.

Unfortunately, some who had been speaking and writing of Standard-Swallows for years took the easy way out with the new name, and continued to use the old one. Worse, sometimes they contented themselves with the

*With an evocative new name and ever-stronger engines beneath their bonnets, Lyons' products were swiftly developing their own identity. The SS Jaguar Drophead Coupe, this one with the 125-bhp 3.5-liter six, typified this fresh blend of style and substance.*

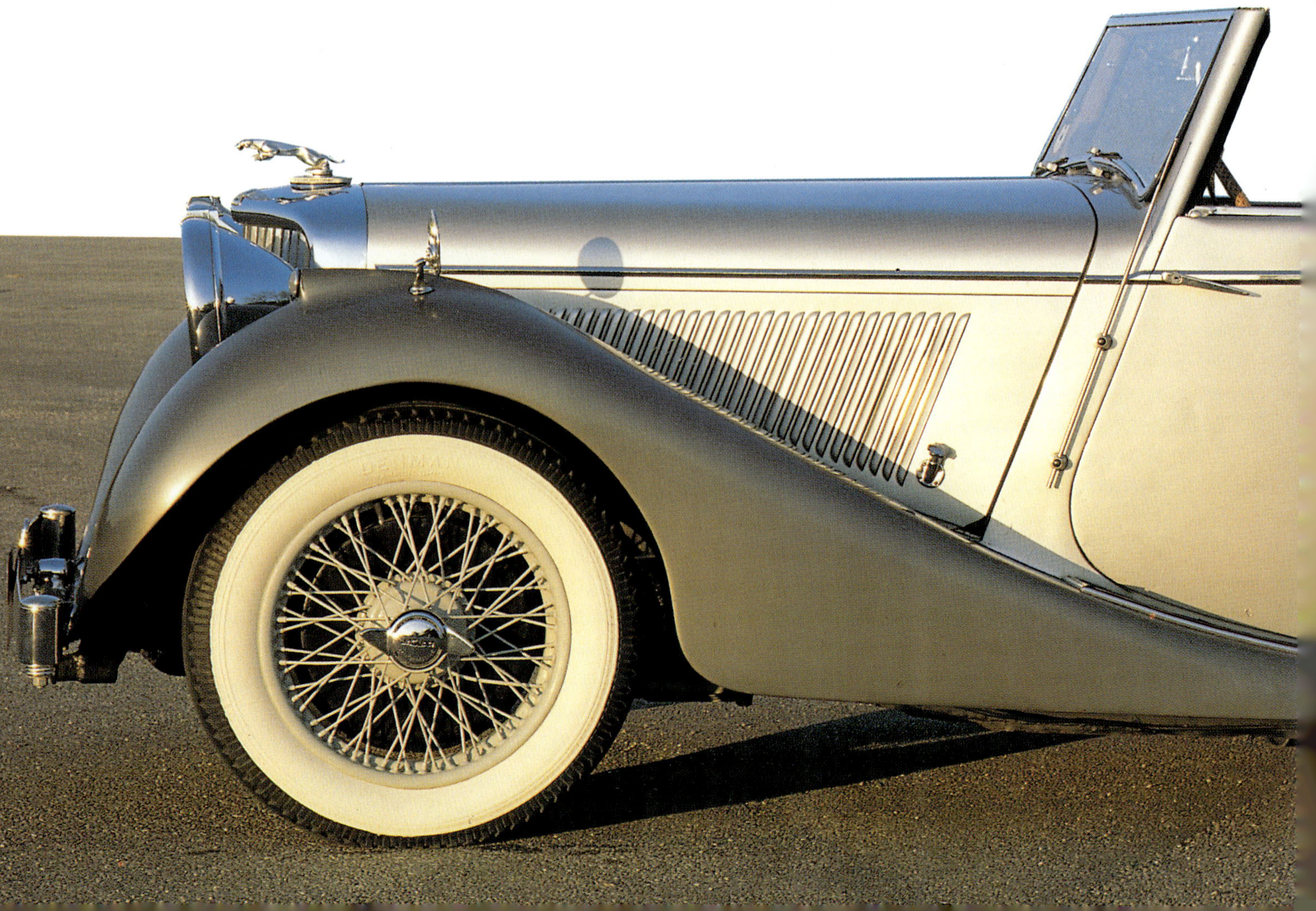

first part alone.

Time to get tough with these people. To launch his new sedan, Lyons determined he must have a new name for it, and asked his advertising people to submit a list of candidates. Among a menagerie of fish, birds, and mammals, "jaguar" leaped out at him.

As he explained, back in the Great War there had been an Armstrong-Siddeley aircraft engine with this name, and it had struck a chord with him. "It had an exciting sound to me."

Perhaps eventually, but not in the beginning. Or so claimed one Bob Bett, quoted by Philip Porter in *The Jaguar Scrapbook*. Bett's brother Bill had started an advertising agency that had Swallow as one of its first clients. Bob joined this firm in 1932, after Swallow had become SS Cars. "SS was a bad name in Europe at the time, with the Nazis being around," he recalled. "My brother decided that they should change [it], and Bill Lyons said, 'All right, come up with some ideas'..." By this account, it was Bob who first suggested Jaguar. "We had a few other names, because we wanted something you could get your teeth into, an animal preferably. But a lot of them were taken up, and Jaguar happened to be available, although it was used by Alvis for an aero-engine."

Bett says Lyons was initially reluctant, perhaps because "he was always a man who wanted other people's views . . . and they were all against it, [believing] SS was a good name." This prompted a lobbying effort by Bill Bett, a two- or three-day stay at the Lyons home in Coventry. Eventually, Lyons agreed to SS Jaguar. "The next day," as Bob recalled, "Bill Lyons phoned up my brother and said, 'I've changed my mind now.' My brother . . . said, 'You're too late. There's an advertisement in tomorrow's *Daily Mail*.' And that was that! My brother really took a chance, because we could have cancelled [the ad], but he was so insistent on it. And, of course, it was the best thing that ever happened to them."

It was certainly an inspired choice, especially given the pronunciation used by most Britishers: "JAG-you-wuh." In sound and image it managed to convey both svelte elegance and primal savagery, a strong yet powerful combination that would work its subtle way on everyone involved with designing the cars, including the boss, not to mention customers. And how appropriate—dare we say inevitable?—that a man named Lyons, born to the land where the King of Beasts had been a symbol of

royal power for centuries, would choose to associate his car with one of the great hunting cats.

That car, which duly went on show at London's Olympia exhibition center, was the new Heynes-engineered SS Jaguar Saloon. Accompanying it was an open-topped SS1, relabeled SS Jaguar Tourer, as well as a development of the SS 90 sports car, the SS Jaguar 100 (usually known simply as SS 100). Reviews were raves in all cases.

That was partly because they offered looks and luxury for surprisingly little money. The new sedan was priced at £385, less than two-thirds what most industry observers had estimated it "should" cost, and the sports car was still under £400. Contemporary rivals of similar performance and style were much more expensive. It certainly helped that additional revisions to the chassis, plus Weslake's new pushrod overhead-valve cylinder head on a bottom end strengthened by Heynes, distanced William Lyons' new marque still further from its humble Standard origins.

That uprated powerplant, a 2.7-liter six now putting out a little over 100 horsepower, was of particular value in the two-seater, which emerged as an able contender in performance rallies and even in certain circuit races. It became even more competitive in 1938, when SS introduced a 3.5-liter enlargement with 125 bhp. That let the sedan break 90 mph and, under ideal conditions, made the "100" a 100-mph car, a magic number at the time. It also put the sports car's price up to £445, but that was still £150 less than the car often regarded as its closest

*By the late 1930s, attaching metal body panels to handmade wooden frames had become too inefficient, so Lyons turned to more-modern all-steel construction. His first all-steel sedan* (both pages) *bowed in '37. It looked much like its predecessor, but was far easier to make.*

direct competitor, the BMW 328.

SS Cars Ltd. operated an owner's club and otherwise encouraged the enthusiasm of its customers, but as a firm it was wary of participating in motorsport. Resources were tight, and in any case the product was gaining such a good reputation without a formal competition program that it seemed pointless to risk public failure. But racing held enough interest for enough key people—including William Lyons, who was known as a fast, competent road driver and who at least once put up best time in a speed test with an SS 100. Thus, the factory undertook some performance development behind the scenes, both for good customers and for its own purposes. One such SS 100 project resulted in an engine of more than 160 bhp and speeds above 120 mph. Both numbers went down in the files and would resurface as goals later on.

Meanwhile, Lyons and his production people had realized it was simply too slow and labor-intensive to continue the traditional "coachbuilt" method of body construction, where metal panels were mounted on a handmade wooden frame. This technique is still considered good enough by Morgan for its iconoclastic British sports cars today, incidentally, but for SS in 1936 it was a bottleneck. The new wave in the industry was all-metal bodies, assembled by welding stampings together, and Lyons had to ride it if he wanted to get anywhere.

Where he very nearly got was nowhere, for there were crippling start-up problems involving late delivery from outside suppliers of stampings that then refused to

*SS Jaguar sedans and Tourers took their model designations from the displacement of their ohv six-cylinder engine. This 1939 Two-and-a-Half-Liter Saloon had the 2663-cc 105-bhp unit good for a top speed of about 85 mph. Engines of 1.5 and 3.5 liters also were offered.*

fit properly. Having hired extra workers to produce planned extra cars, the company came dangerously close to the financial brink in 1937, but at the last minute Lyons and his people managed to get things together. Looking very similar to the four-door introduced two years earlier but in fact substantially revised, the new "all-steel" sedan was roomier, lighter, easier to make, and, in the end, more profitable. In the fiscal year ending July 1939, a company workforce of more than 1000 shipped out more than 5000 cars.

Just what Lyons might have produced next will never be known, for war intervened. Late in 1939, SS Cars packed its automotive tooling away and began fulfilling military contracts. Now its recently gained expertise in all-metal bodywork came in handy, for much of its war work involved sheetmetal parts for aircraft—fuselage sections, bomb bay doors, wingtips, drop tanks. The benefits ran both ways, however, for such projects brought the company fresh knowledge, advanced techniques, and modern machine tools that would be useful in peacetime.

One automotive project was undertaken during the war. Trying to develop a jeep-type vehicle small and light enough to be dropped onto battlefields by contemporary parachutes, the factory made up a pair of prototypes with all-independent suspension and stressed-skin monocoque (unitized) construction. The project came to nothing, but this experience, too, would prove valuable down the road.

Typical of the man, Lyons made the best of the war years on the management level. It was policy that everyone take a weekly turn at manning the factory against the bombing raids that so devastated Coventry, and he organized schedules so that he and his top engineers all drew the same fire watch on Sunday nights. They'd occupy the quiet times by planning their moves once the war ended. It was during these long, congenial hours that the great Jaguar XK engine was conceived. Lyons was not an engine expert himself, but he was firm that the new powerplant should offer genuine high performance, and that it should look the part, too. If he was going to put the Jaguar name on an engine, it was going to be an engine worth looking at.

Building his own engine was now within Lyons' scope, for he'd purchased the appropriate six-cylinder tooling from Standard. Bit by bit, his old dream of becoming a "complete" automaker was coming into place.

It was in this mood that Lyons finally let go of the last vestige of Swallow—and a moniker that would have been woefully out of place in the postwar world. While impatiently awaiting the war's end, he changed his company's official name from SS Cars to Jaguar Cars, Ltd. in March of 1945. Six months later, as if to celebrate the return of peace, he introduced the first cars to carry no name but Jaguar. But in virtually every other respect, these were no more than the prewar models reprised. The biggest difference was the absence of "SS," a name public opinion simply wouldn't have tolerated in the wake of Nazi atrocities.

Of course, the war had changed far more than the connotation of certain initials. Britain itself was a different

*Repeated World War II German bombings devastated central Coventry* (far left, bottom), *but only once did they seriously damage the SS Cars' factory itself. Lyons dropped the war-tainted SS prefix in '45, rebadging his cars, simply, Jaguar* (opposite, top).

*SS Cars' military work included fuselage sections* (top right) *and the development of the jeep-like VA* (above) *and VB* (right). *Experience with the experimental vehicles' lightweight unitized construction and independent suspension proved useful after the war.*

place, and a sluggish economy plus severely rationed fuel made it a poorer place for automobile sales in the immediate postwar period. Urged on by government incentives involving the supply of steel, Jaguar began a serious export campaign. The first left-hand-drive models came along, and convertibles were reintroduced. Early in 1948, Lyons spent five weeks touring the United States, setting up distributorships. While there, he managed something of a publicity triumph by being on hand when movie idol Clark Gable took delivery of the first Jaguar sold in Hollywood—a convertible, of course. Soon, the bulk of the company's output was earning foreign currency, and Lyons was getting all the steel he could use.

Behind the scenes, William Heynes and his staff had been working hard on several all-new models. The first of these would be a sedan, the company's proven bread-and-butter product. Though it would honor contemporary Jaguar traditions as to appearance, it would embrace important modern technology: hydraulic brakes and independent front suspension. This vehicle was first unveiled at the end of September 1948 with a model designation new to Jaguar: Mark V ("Five").

Although it was not part of the Mark V specification, there was an all-new powerplant coming along, too: the six-cylinder, twin-camshaft XK engine conceived during those Sunday-night brainstorming sessions during the war. It was being refined for still another sedan, a genuinely modern conception that would debut in 1950 as the Mark VII. ("Six" would be skipped because Bentley was using it.)

But the new engine was ready now. Why not let the world see it, let it prove itself in the public eye, demonstrate that Jaguar was really forging ahead? And why not keep faith with the cadre of SS 100 owners who

*Lyons had a bombshell of his own in the XK 120. It debuted in October 1948, its new engine a masterpiece, its body a sensation. Orders pouring in from overseas markets gave postwar Britain a vital infusion of cash and an immeasurable dose of pride. American actor Clark Gable was a fan* (right). *The XK 140* (above) *followed in '54 and improved on the successful XK 120 formula.*

*Sports cars got headlines, but sedans and coupes were Jaguar's backbone. Introduced in September, '48, the Mark V, seen above in Drophead Coupe form, was Lyons' first new postwar car, though, unlike the XK 120, its engine was prewar.*

were still using the prewar cars in competitions, but who were most sensitive to the rapid postwar developments in their field of enthusiasm? One of the most successful of them, as it happened, may have enjoyed a special conduit to William Lyons: Ian Appleyard, a rallyist who was going to marry his daughter, Patricia.

Out of such influences was born the immortal XK 120, displayed for the first time in late October of 1948. Its birth was hurried; Lyons later recalled spending no more than two weeks knocking up the styling buck. And he initially planned to spend little more time manufacturing these "Super Sports," as he initially called them (perhaps surreptitiously testing the viability of the old SS abbreviation). There were only going to be about 200 of them, and for such low-production items the old-style coachwork construction would be adequate. These original XK 120s were clothed in aluminum panels hand-assembled over wooden body framing, all mounted atop a shortened Mark V chassis.

But the stunningly beautiful, astonishingly fast machine ignited public enthusiasm to rare incandescence. Lyons realized he simply had to tool up for mass production. It took a painfully long time, a year-and-a-half, to get the line going. But doing so was probably the single act that best established the stature and longevity of his company. It is no exaggeration to say that the XK 120 was primarily responsible for making the name Jaguar legendary.

*The Mark V* (above) *carried on Jaguar's prewar design themes and its Standard-based 2.5-and 3.5-liter sixes dated from the '30s. But a new, very rigid chassis and modern independent front suspension made it quite adept on the road.*

With 160 bhp and 120-plus mph, the new roadgoing sports car offered the same performance as the prewar SS 100 development racer. It was thus inevitable (there's that word again) that the XK 120 itself would be raced. With its normal reservations assuaged by the car's potential, the factory launched the 120's speed career. First, in May 1949, came a maximum-velocity demonstration on a closed-off section of long, arrow-straight superhighway in Belgium—Britain then lacking venues of that description. Under the opened eyes of press flown in for the occasion, a stripped-down roadster topped 132 mph. Then, that August, Jaguar entered three cars in a one-hour contest at the Silverstone track in England. The same XK 120 used in Belgium won the model's first-ever road race.

Thus encouraged, the factory decided to try a trio of 120s in the 1950 Le Mans 24-hour enduro. One actually reached second place at one stage before its clutch failed, but the other two kept going in midfield all the way to the finish line.

With this began Jaguar's glorious run at the French classic. So exciting was such a result with what amounted to standard road vehicles that Lyons gave his blessing for a full-on effort with a special "competition-type" XK 120 for 1951. This C-type Jaguar won the world's most important manufacturer's contest the first time out and its magnificent XK engine would go on to do the job four more times.

Meanwhile, Appleyard had finally gotten his

*Unveiled in 1950, the Mark VII sedan had not only the hot XK twin-cam six, but its flanks echoed the sports car's cat's-paw fender line. It was a hit in the showroom and a surprising success in all manner of competition.* At left, *a Mark VIIM of '54 in mid rally.* Below, *the little-changed Mark VIII iteration of '56—here mysteriously wearing a Mark VII factory plate.*

hands on a 120, and with new wife Pat along as navigator, won the 1950 Alpine Cup. It was to be the first of four of those for him, and of numerous other rally trophies.

These were only a very few of a multitude of similar victories for the sleek new cat, and it soon emerged as the most important car Lyons had ever created. In fact, every Jaguar for the next 25 years owed its genesis to the XK 120. In 1950 the Mark VII sedan came out with not only the same engine and basic chassis concept but also the same overall styling theme, and was popular enough to retain these through two revisions, the Marks VIII and IX.

Jaguar also put the magic 120 formula into a series of smaller sedans (ultimately called "Mark II") with new-technology monocoque chassis, beginning in 1955 with one called the 2.4-Litre, for its destroked XK engine. That was later joined by a hot 3.8-liter that would win fame as both a racer and a police-pursuit vehicle before being retired a dozen years later. Even then, the basic structure with new body lines continued on for several more years in cars called S-type and 420. There were also all-new sedans in this period—the large Mark X ("Ten") of 1961 and the more compact XJ-6 of 1968—but they were still powered by the wonderful Jaguar XK engine, and continued to make more sedate use of chassis elements developed for the sports cars.

As for those two-seaters, the XK 120 was developed into 1954's XK 140 and on into the XK 150 three

*In February 1957, a fire in Jaguar's Browns Lane plant destroyed scores of cars under construction, as well as valuable jigs and tooling. Some parts were salvaged* (right), *but the biggest loss was the raw materials for production of the tantalizing XK-SS sports car* (above).

years later. By that time, admittedly, the basic design was aging, but it is interesting to note that the XK-series was several times selected by specialist body builders to carry their own styling exercises. How the Jaguar founder must have smiled to see how far he'd come.

William Heynes and his staff surely were pleased, too, at the number of small-scale constructors who had chosen their engine to power racing cars. The twin-cam Jaguar six had taken on in its homeland the performance importance that Chevrolet's small-block V-8 had in America, and circuits all over Great Britain were tingling to the rasping exhaust note of HWM-Jaguars, Cooper-Jaguars, Tojeiro-Jaguars, and Lister-Jaguars.

The factory's own C-type racer, meanwhile, having again won Le Mans in 1953, gave way under the pressure of racing to the D-type. This used the same mechanical elements in a very advanced chassis and body to triumph again in '55, '56, and '57.

The D-type Jaguar was a racing car, but it had been built to regulations that kept it close enough to road-going sports cars so that Jaguar could turn a few leftover D-types into a street model named XK-SS—those mystic letters being floated all at once. Partly due to a nearly disastrous factory fire in 1957, but more to its unarguable impracticality, that particular model had a short life, but it was the direct ancestor of William Lyons' second great sports car triumph: the E-type.

Sprung on his enthralled public in 1961, this was a functional, docile, even luxurious everyday vehicle that held honest racing credentials (like the 120, it won its first-ever race), and which had sensational looks besides.

An irresistible blend of sex appeal and performance with practicality, the XK-E, as it was known in America, was epochal in its impact. Very few automakers ever have scored once as Jaguar did with both its XK 120 and its XK-E.

As early as 1950, Jaguar was putting out so many cars that its original Coventry home at Foleshill was feeling cramped. Having already expanded its original floor space by a factor of 15, from 40,000 square feet in 1928 to some 600,000 now, the company applied for permission to build more. The local authorities refused. So the decisive Lyons went shopping, and during 1951 and '52 moved his entire, massive operation bodily to an existing facility only two miles away, but in another district. Off a street known as Browns Lane, running north from the suburb of Allesley, the new buildings enclosed a total of one million square feet. This remains the company's main plant today, although the firm has since expanded into three more sites around Coventry.

At the end of 1955, for his achievements in national prestige, in international style, and not least in foreign exchange, Lyons was rewarded royally: named in the New Year Honours List as a Knight Bachelor. The new Sir William was soon playing host to the Royals themselves, conducting Queen Elizabeth II and Prince Philip on a tour of Browns Lane in March 1956. The visit was well-publicized, of course, and Coventry city officials took care to mark the occasion by presenting Her Majesty with a beautiful scale model of the city's most prominent product, the Le Mans-winning Jaguar D-type. "Billy" Lyons,

*Jaguar found it too expensive to restart the XK-SS line after the fire, so assembly of the two-seater ended with fewer than 20 built. Enormous effort enabled the plant to resume production of other models within days, however.*

one-time motorcycle racer and car salesman drifting through life, could scarcely have dreamed of scaling such heights.

Tragically, he was not standing at the pinnacle of personal happiness. The spring before, his only son, John, had died in a highway accident on a trip to Le Mans.

It is the most idle of speculation, but Jaguar's subsequent history might have run a different course had there remained an obvious heir to its ownership. Now, in eerie company with Enzo Ferrari, who had lost his son Dino that same year, William Lyons would one day be forced to consider other forms of succession.

His sorrow did not sink him. Lyons did withdraw the factory from direct racing involvement at the end of the 1956 season, leaving it to private teams to carry the flag. But this was done so that he could devote all his resources to the production side—and to building his firm still bigger. In 1960, once again needing more manufacturing space, he purchased Daimler, the wholly British firm that had begun as a licensee of the pioneering German automaker early in the century. Contrary to fears at the time, Lyons did keep that honorable old marque in production, even retaining its mid-size sports car, the SP250, at least through 1964, along with a well-regarded line of commercial buses. He also adopted Daimler's nice little V-8 engine for some of his own cars; these, and others down the road,

*By the mid '50s, Jaguar had secured a special place in the English heart. Perhaps Britons had come to believe that these cars, cultured and often indomitable, reflected the best in their nature.* Below left: *an XK 150 of '57.* Bottom: *the 2.4 Saloon of '55.*

*In late 1955, for his achievements in national prestige, international style, and foreign exchange, William Lyons was rewarded with knighthood.*

were given different detail styling and called Daimlers. Working in the opposite direction, a large Daimler limousine was developed that used numerous Jaguar components, including the venerable XK engine. This largely hand-built carriage remains in production today.

Continuing his expansion, Lyons took Jaguar into areas that surprised some observers. For example, he entered the heavy-truck field in 1961 by acquiring Guy Motors, whose product line continued for another 18 years, long past the time when Sir William had relinquished control of his empire. Expanding the portfolio further in 1963 was Coventry-Climax, the fine old firm famous for championship racing engines as well as hand-transportable fire pumps, though what attracted Lyons was the high profitability of C-C's forklifts.

The Jaguar Group continued to grow and prosper until that day in July 1966, only a few weeks short of his sixty-fifth birthday, when Lyons announced that he had come to a merger arrangement with the British Motor Corporation. He would not be stepping down right away, he said; indeed, he would continue in an active consultancy role for several more years. But anyone who had come to love his cars feared the worst immediately.

The company's entire body and soul had been the life's work of one strong, gifted man. What would happen to Jaguar with Sir William gone?

*Lyons' vision included an XK engine versatile enough to serve an astonishing variety of machines, from sedans, to the HWM-Jaguar racer* (above and right), *to the libidinous E-type sports car* (below).

CHAPTER IV

# JAGUAR AFTER LYONS

We are judged in history not only by our accomplishments, but also by the void we leave behind. Sad to say, Sir William Lyons' departure from day-to-day control of his company in the mid sixties revealed him as a great man.

Sad, because Jaguar Cars very nearly failed to outlive its creator. Granted, the times were hard. Carmaking itself had changed enormously in the four decades since Lyons had become involved, and as he faced retirement he realized his country's whole economic climate was turning increasingly harsh. Problems stemming from labor unrest, governmental misdirection and energy shortages afflicted every business in Britain. Small operations such as Jaguar were threatened by overwhelming new costs of developing fresh product, as well as by stiffening competition from foreign automakers. Together, these factors were fostering almost a mania for merger throughout the industry. What's more, Lyons' only heir had died many years before, and his extremely self-reliant management style had never nurtured a successor within

*Jaguar's 1990 Le Mans-winning XJR-12 No. 3 and the specially tuned '90 JaguarSport 6.0-liter V-12 XJR-S Coupe* (left in photo) *and 4.0-liter six-cylinder XJR Saloon symbolize a new era at Coventry. The faithful see in such cars a renewed sense of the aggressiveness that made Jaguar so great under William Lyons.*

the company.

So, all in all, liaison with the British Motor Corporation (BMC)—which just happened to own the body-panel plant on which Jaguar production absolutely depended—may indeed have been the best course across the difficult ground he saw ahead.

But it led to near-calamity. What Sir William saw as a merger, his new associates considered a purchase. A successive takeover by another group soon afterward turned the tables, in a way, but left Lyons with even less authority. Finally absent its master's hand, Jaguar lost its way, and began a long stumble into darkness. Despite what one must suppose were good intentions, the path paved ran downhill. For many dreary and turbulent years, the once-proud specialty manufacturer subsisted as an orphaned stepchild, unwanted and unappreciated, misunderstood and misused. Eventually the firm fell into the possession of Britain's then-socialist government, under whose deadening hand efficiency, initiative, quality, reputation, sales, spirit, all sank ever lower. There came a somber hour when there loomed a very real risk of corporate extinction.

Astonishingly, against all likely odds, Jaguar survived. Survived, recovered its health, renewed its prosperity, regained its independence, resumed the production of cars worth their buyers' admiration. When the little factory was again taken over, by Ford Motor Company in 1990, it fetched the stunning price of 2.5 billion U.S. dollars.

How it all happened is one of the industrial epics of the century, and rightly has been recounted and celebrated in many formal studies. Here, it is enough to sketch the outline; frankly, the period of deepest twilight produced little of interest to the enthusiast.

Jaguar's merger with—or sale to—BMC, a full-spectrum automaker known in America primarily for its

*Ford Motor Company bought Jaguar in 1989, but appears quite aware of the importance of letting the proud cat continue to leap to a British tune.*

Austin-Healey and MG sports cars, created a new entity called British Motor Holdings (BMH). The arrangement was announced in July of 1966. A year-and-a-half later, in January 1968, Lyons, now aged 67, stepped down as Jaguar's Managing Director. His old race team manager, Lofty England, partly replaced him as one of a pair of joint MDs. Sir William, having held onto the titles of Chairman and Chief Executive, continued to work toward his last pet project, the XJ6 sedan. The BMH bosses may have considered him their property, but they recognized both his value and his stature, and were happy to let Lyons continue in his own old ways.

So far, so good, but a few months later occurred a further corporate condensation: Leyland, builder mainly of commercial vehicles as well as Triumph and Rover cars, muscled itself into a merger with BMH. The new conglomerate, formed in May 1968, was known as British Leyland Motor Corporation (BL). Though considerable lip-service was paid to Jaguar's importance within the new scheme, there was also plenty of talk about consolidation, rationalization, and production economies.

Obviously, this threatened further submersion of Jaguar's identity, and was not what Lyons had visualized two years earlier. Gamely, he put a good public face on it. He agreed that increasing competition from overseas forced a certain unity on the British industry, and pointed out that as a continuing member of the board, he still would work to preserve the autonomy he regarded as essential to Jaguar's special market appeal.

In any case, as an astute businessman, Lyons recognized that a need for consolidation, rationalization, and production economies had reached even Jaguar's own product range. Out of the Browns Lane factory were rolling not only the E-type sports car in three different configurations (the new 2+2 in addition to the well-established open and closed two-seaters), but also a bewildering array of sedans. These included a development of the big Mark X, known as the 420G, as well as several permutations of the old mid-size Mark II: the 420, the 340, the 240, all with six-cylinder engines, plus one with a small Daimler V-8. Some of these basic cars had first been seen back in 1955. With his new car, Lyons was determined not only that Jaguar would leap into the seventies technologically, but also economically—the one model would assume most of the market roles previously played by the multitude.

Unveiled in September 1968, the XJ6 has often been characterized as Lyons' best work. As marque historian Andrew Whyte put it, the graceful new four-door was the "ultimate Jaguar motor car . . . intended to be the car with the world's quietest and most refined riding qualities, bar none, and [to] retain Jaguar's famous attributes of performance, controllability, comfort, and value for money." The XJ6, he continued, "really *was* the culmination of all the ambitions Sir William Lyons had for the company he had founded. . . ."

The new model was well received, even winning car-of-the-year awards. There arose bedeviling glitches on the production side, but still Jaguar Cars remained a money-maker—one of BL's few profitable divisions. Those in overall charge felt ever-increasing temptation to exploit the image. They seized their chance when Lyons finally retired altogether in 1972.

That was Jaguar's "Jubilee Year," its fiftieth since William Lyons and Walmsley had set up in the motorcycle sidecar business. Sir William celebrated his company's official birthday—and his own seventy-first—on the fourth of September. Within weeks, Leyland was moving openly to stir its one golden goose into the leaden stewpot of its lesser brands. Jaguars began to lose their exclusive aura as the public found them displayed and sold alongside plebeian everyday conveyances. Worse, the ardently loyal Jaguar workforce discovered that Leyland planned to move vital operations to other plants. There was even talk of cheap future Jaguars that would be Jaguars in badging only.

Labor turmoil cropped up, too. In mid-1972, much of Jaguar's workforce walked out for ten weeks. Sir William voiced "deep concern" for the company's future.

Indeed, it looked bleak. So many of the original Jaguar people were gone now. William Heynes had retired in 1969. Replacing him was Walter Hassan, a good man and long-time Jaguar stalwart, but himself not far from retirement. Lofty England was now Chairman and Chief Executive; he was 60.

To bring in new blood, in 1973 the vast BL organization appointed one of its own "bright young sparks" to be England's nominal understudy and eventual replacement. But the 34-year-old Geoffrey Robinson very soon made it very clear that he had no intention of serving any period of apprenticeship, and launched an ambitious—and expensive—program of modernization, expansion and production increases. England, feeling left out of things, took early retirement in 1974. By the end of that same year, a year of energy crisis that forced three-day work weeks in Britain, the entire British Leyland operation was so deeply in debt that the national government became alarmed. A commission was formed, headed by Sir Don Ryder, to look into the mess. In April 1975, the commission unveiled its so-called Ryder Plan. This essentially reversed Robinson's course (he promptly quit), and put the nearly bankrupt Jaguar under government control within a new entity called Leyland Cars.

That didn't work either. By the end of 1977 the Ryder Plan had been discredited as unrealistic and unworkable (Ryder promptly quit). The struggling nationalized firm tried something else. Under the direction of a new strongman, Michael Edwardes, the BL name moved to the background, the Leyland label went back on commercial vehicles alone, and the passenger-car line split in two: Austin Morris to make bread-and-butter sedans, and Jaguar Rover Triumph (JRT) handling more upscale products.

But in another two years, the JRT scheme had

*Lyons stepped down as Jaguar managing director in '68, but continued work on his pet, the XJ6 sedan* (left). *Lofty England* (above) *managed Lyons' race team in the '50s and was a Jaguar CEO in the '70s.*

proved itself a failure too (though Edwardes was granted a knighthood). There was too much difference between the products, and too many differences among the personnel. Another strike came in April 1980. Management issued ultimatums, and rumors spread of Jaguar being removed from Browns Lane entirely. It looked as though the Coventry Cat had used up its last life.

Enter John Egan. A vigorous 40, he came from a town called Rossendale in Lancashire. That happened to be the same county from which Lyons had sprung, and the two are said to have shared many of the same traits of character, notably, a certain straightforward, no-nonsense, even blunt approach to problems. Furthermore, Egan had been educated in Coventry, so he had strong roots in Jaguar's hometown. His own career rested on a foundation of engineering bolstered by a business degree, plus management experience in the international arena, much of it with BL. An obviously able man, and in fact he'd already been approached to join Jaguar some years earlier. He had turned down that invitation, but early in 1980 Sir Michael Edwardes finally persuaded Egan to move into the old Lyons office.

The timing was terrifying. The assembly lines were immobilized by the strike. As he would joke years later, Egan started his new job wondering if "I was destined to be the only head of an automobile factory who never built a car."

But, ignoring hints from his superiors that he need not get involved in the labor mess, Egan devoted his first weekend to intense meetings with the militant shop stewards and managers. On Monday morning, they reported for work. Clearly, the plain-speaking Lancastrian was an inspiration, and in the following weeks he made good on his promises. Showing a deep dedication to Jaguar and its revitalization, Egan demonstrated such perseverance, enthusiasm, and willingness to listen to any complaint that the entire plant caught his spirit. Acknowledging the hard truth that quality had slipped badly, he made that his first priority. He instituted Japanese-style discussion groups, moved to better integrate the badly scattered supply network, and boldly addressed massive body-assembly problems. And every night he took a new car home to check it out.

Egan did enjoy the vital advantage that many of his staff were hard-core company partisans who themselves had fought a long guerilla war to keep Jaguar, Jaguar. One man in particular has often been called a special hero: Bob Knight, who had succeeded Hassan as director of engineering in 1972. While Leyland had absorbed most of Jaguar's formerly autonomous departments, such as sales and service, Knight simply refused to let his go. Through sheer British bulldoggery, he kept engineering clenched tight in his own jaws, and in Coventry. It was chiefly thanks to him, says everyone who knew the story, that Jaguars never became Rovers, or Austins; that they remained definitively Jaguars.

But nobody at the revitalized Browns Lane plant had any illusions. They all knew who owned them. They

*Jaguar heroes: John Egan, pictured at left with his wife, revitalized management and production, while Bob Knight* (above) *fiercely defended the autonomy of Jaguar engineering.*

all knew that such historic British flag-carriers as Austin-Healey, MG, and Triumph already had been killed off as hopeless, and that Jaguar was being scrutinized by beady eyes. "If we had failed," John Egan has remarked flatly, "Michael Edwardes would have shut us down."

They didn't fail. They improved the product. They raised productivity. They increased profitability. They purged the dealership structure of marginal operations. They revitalized the reputation, and rekindled the confidence. They saved Jaguar.

Although seemingly a peripheral activity, racing became an important element in the overall success. That first year, 1980, Egan recognized the positive contributions being made to Jaguar's performance image by Bob Tullius' Group 44 racing team in America, and pledged renewed cooperation. By 1982 he'd become impressed by Britain's own Tom Walkinshaw Racing, and authorized factory support. The payoff included not only numerous victories all over the world with both modified XJ-S coupes and XJR-series racers, but eventually the grandest prize in sports-car competition, the 24-hour race at Le Mans, Jaguar's old field of special glory. The value to the company of all this performance prowess was indirect, but valid: It demonstrated the merits not so much of the product, as of the people and their renewed spirit of competitive determination.

Not coincidentally, the company's return to commercial success paralleled a profound redirection of Britain's leadership. In 1979, the voters had turned the Labor party out of office, and the new Conservative government immediately set about putting various nationalized industries back into the hands of private enterprise. In preparation, Jaguar's custodians spent years diligently building up its value. Further government investments went into modern robotic assembly equipment, and also seeded a new generation of XJ6, the XJ40, with an all-new engine (AJ6) to power it. Profitability soared; whereas in 1980 the factory had managed to turn out only 1.4 cars per employee and posted a loss of £47.3 million, in 1983 productivity more than doubled to 3.4 cars per worker and the books showed a profit of £50 million. The following year's numbers were better yet, and in August 1984 Jaguar stock was offered on the open market. The response was tremendous. London's *Financial Times* likened the frenzy to "the start of a sale at Harrod's china department." Even Sir William bought.

Just short of his eighty-third birthday then, the founder was still very alert, and keenly interested in the fortunes of his old firm, which he still served as Honorary President. During a visit to the Lyons estate, Wappenbury Hall, south of Coventry the following January, author Whyte found a man "very much 'on form', wanting quick answers to his questions—as always—and expressing his satisfaction that, at last, his own record for a year's deliveries of new cars had been exceeded." Only a few weeks later, on the eighth of February, 1985, William Lyons died.

He had been one of the grand figures of the

*Sir William Lyons poses with a depiction of one of Coventry's proudest moments: the C-type roaring to Jaguar's first Le Mans win in '51. Lyons died in 1985, at age 83. His legacy lives on in the company's R&D center at Whitley and in the fast, refined XJ-S* (above)

*Handworked hardwoods are still a part of that Lyons legacy. And today they combine with some automated assembly to create modern Jaguars.*

world's automobile industry, and he had lived to see his life's work succeed. However, success breeds not only success, but insecurity. More than ever, Jaguar was now a very attractive takeover target. Through the last of the eighties, the company was sheltered from unwelcome acquisition by something called the "Golden Share," basically a government regulation that limited any one stockholder to 15-percent ownership. But that arrangement, put in place when the company was privatized in 1984, was due to run out in 1990. Jaguar would become fair game.

International game. BMW, Ford, and General Motors had all been sniffing around the Browns Lane doorstep for years. To the staunchest of the Brits, the prospect of foreign ownership of something so deeply British as Jaguar was anathema. Unfortunately, no local firm had the resources to buy what had become a very valuable property. After all, the bulk of the native auto industry already was in overseas hands—Chrysler had long since acquired Rootes, GM already owned Vauxhall and Lotus, while Ford, long a strong player in England under its own name, recently had bought Aston Martin. Honda had a 20-percent stake in Rover.

Now titled Sir John Egan, the Jaguar chairman fought to keep the company independent. But he was far too aware that as a small private firm, Jaguar simply could not command the resources required for increasingly vital research and development of upcoming models. Sales in the U.S. market, always important, had recently softened. For months he courted GM, hoping to bring in the necessary funding in return for only a minority share. But Ford, flush with cash from a decade of success at home, was prepared to pay—to pay astonishingly well: $2.56 billion—for the whole package. Bowing to the inevitable, and saying it was thinking of the good of the company, the government canceled the Golden Share regulations a few months early.

When the announcement came in November 1989, Egan explained why he'd finally changed his mind. "Independence is one matter; the shareholders' requirements is another matter, and also, looking at the requirements for finance and technology for the future is another one." Acquisition by Ford, he added, was "a better way forward for both companies." And at least Ford did promise to preserve Jaguar's uniquely British identity.

What did Ford buy? Image and prestige, of course, as well as considerable engineering expertise. But mainly Ford purchased a quick, relatively painless entrée into a luxury market segment that was growing, but in which it had no ready representation.

And what did Jaguar fans get? A revitalized, well-nourished, healthy Cat. A specialty automaker able, once more, to develop and manufacture the sorts of exciting new cars that had built the name into a true marque—one of the best-loved in automotive history. Sir William's great legacy appears safe.

CHAPTER V

# SS JAGUAR 100: DISCORD AND HARMONY

Extravagant, indeed outrageous, it assaults the eye, commanding attention, its dissonances refusing to be ignored. How solid the long, muscular body, crouched between the wheels like a lioness poised to spring; yet how ethereal the four fenders, swooping gracefully around this latent fury like taunting swallows. The SS 100, a compelling commingling of power with grace, of the bestial with the beauteous. Extreme discords, uniting in

exquisite harmony.

It was with this car that William Lyons made his company. Not in numbers, no, but surely in image, in mystique. It appeared at the time SS Cars was adopting the name Jaguar, and what a splendid great predatory feline this car looked to be.

Yet had it claws? Was there sinew below the skin? If not, this dazzling coachmaker's confection would fade into darkness. Promise is not enough in the performance business.

But the SS 100 amply fulfilled its promise. If not the greatest sports car of its age, it was good enough to be seen as a full-blooded one, and is thus the progenitor of every sporting Jaguar built since. With it, Lyons compounded the explosive formula that would catapult his firm to success through the next half-century and beyond.

Lyons and his early partner, William Walmsley, had started as automakers by "packaging," a 1920s British term for replacing original factory bodywork with more stylish shapes, brighter colors, and more sumptuous trim. Thus offering what we might now call "image enhancement," they found a substantial market immediately, and their business grew rapidly.

But Lyons had farther sight. Out of his personal ambition, and probably also out of sheer survival instinct, came his decision that Swallow Cars, Ltd., would have to move on from packaging, becoming one day a true "tires-up" automobile manufacturer.

He took the first step down that long and winding road in 1931, when he persuaded his major supplier, the Standard Motor Car Co., Ltd., to make up some modified chassis for Swallow's exclusive use under a new car he would call the SS1. The main alteration to Standard's Essex frame was at the rear, where the leaf springs were located outboard of the chassis rails, rather than under. With the adoption of flat, rather than arched, springs front and rear, this allowed Swallow's new body to be carried dramatically low—the roofline was a startling 13 inches lower than the standard Standard's. As one enchanted newspaper writer put it, "Two short people can shake hands over the top." Obviously a new experience in saloon-car motoring.

Power naturally came from Standard's inline six, a side-valve unit of insipid performance but robust, seven-main-bearing construction. This was offered as a basic 2054-cc unit (125.3 cubic inches) and a slightly more potent 2551.4-cc version (155.7 cid) in respective models called Sixteen and Twenty. These names reflected taxable horsepower, which was lower than actual. Not that actual horsepower was high, but the engines mounted well back under a long, long "bonnet" for the *look* of power.

Longer-and-lower is a fetish usually ascribed to Detroit in later years, but it was in fact Lyons' aim with this new model. He meant to attract attention, to make a statement.

He succeeded. Though the first SS1 of model year 1932 was completed in haste and actually displeased Lyons, it was a hit with the press and a success with the public, which bought all 500 of his theatrical *oeuvres*.

By the standards of its day, the SS1 was considered to have good road manners— "taut and correct," was one judgement. It also had interior room and outward visibility that surprised people who had formed preconceptions from the exterior appearance. It was well-finished inside, of course, with leather and wood trim, comprehensive instrumentation that included an electric clock, and the lady's vanity set with mirror that had been a special Swallow feature from the beginning of its bodymaking.

Performance, however, didn't really match the implied promise of the styling. Even the larger Twenty engine produced only about 55 horsepower. Admittedly, speed doesn't seem to have been a vital concern to most

*Only 308 were produced over three years, but the SS 100* (both pages) *was seminal in design and performance. It also was a smashing commercial success for a small company striving to make its mark. With the SS 100 and its close companion, the SS 90, William Lyons struck the pattern for every sporting Jaguar to come.*

SS1 customers, as the smaller (and less heavily taxed) 45-horse Sixteen engine proved more popular.

For 1933, the SS1 design was substantially revised. For one thing, Standard pumped up both engines a little by means of higher compression and better carburetion. The 2.6 engine now produced about 62 bhp. More significant were Lyons' chassis and body alterations. He got Standard to widen the track by two inches, to 51, which would give both more interior room and more stable handling. He had the chassis itself changed again. It was made stronger, its rails were run under the rear axle for lower body height, and the wheelbase was lengthened by a substantial seven inches, to 119.

Though the revamped second-series SS1 remained more of a 2+2 than a sedan, its deeper, longer cab gave more room to rear passengers. It also brought the original car's enormously long hood into better aesthetic balance.

Lyons seized the opportunity to redraw the side windows, so they now harmonized with the body's beltline. He also lightened the rather heavy, Germanic character of the original car by deleting the "helmet," or cycle-type, front fenders in favor of long "clamshells." These swept gracefully back and down through newly added running boards to unite with the rear fenders, giving the profile a unity it had lacked. While at it, he reshaped the radiator shell and grille, deleted a brace bar between the headlights, and slanted the many louvers in the sides of the engine hood to match the door openings. Altogether, the effect was of a much more refined, self-possessed, mature automobile.

This second-series SS1 sold even better, prompting the eventual development of three additional body styles, all two-doors. The initial Coupe, which kept its backseaters pretty much in the dark, was joined by a "four-light" Saloon that gave them their own side windows. There was also a more faddish "Airline" version of this body, a square-roofed fastback whose windowline drooped down to a rounded tail. Customers who craved fully topless motoring could opt for a convertible, which bowed as the Open Four-Seater Sports. Later sold simply as the Tourer, it was the genesis of the SS sports car.

Although the company didn't really intend the convertible for serious sporting use, it became quite popular with customers who enjoyed "driving tests," something like modern gymkhanas, and long-distance "trials,"

*Lyons grabbed attention with his low-slung SS1 and its impossibly long hood. The early factory rendering above actually shows a lower roofline and a higher bonnet profile than the more-awkward 1932 production car's. The chassis and six-cylinder engines were by Standard Motor Car. Revised for '33, the SS1* (opposite, top) *used the same mechanicals, but got more power and a longer wheelbase, while swept-back fenders and slanted hood vents helped add a dash of grace. The "Airline" Saloon body style of '34* (opposite, bottom) *catered to the fashion of the day.*

or rallies. Perhaps to the annoyance of some critics, who never could understand "how they do it at the price" and assumed shoddy materials and workmanship under the surface gloss, the SS1 proved itself sturdy and reliable over the often atrocious European road surfaces of the 1930s.

Keeping faith with its fans, SS Cars further upgraded the basic car for the 1934 sales season. The body stayed substantially the same, but the chassis' central X-member was moved forward slightly for more rear foot room, and the track was widened by another two inches, to 53.

The biggest news was up front, where engines were larger, though not by much. The smaller unit grew by a mere 89 cc (5.4 cid) to 2143 (130.7 cid); the more historically important 2551.4 expanded to 2661.9 cc (162.4 cid) via a longer stroke (106 mm versus 101.6) on an unchanged bore (73 mm). Breathing was again improved. Previously, a single carburetor on the right side fed intake valves on the left through a block passage between the middle cylinders; now, the carb was more efficiently mounted on the valves-side. With all this, the new 2.7 delivered about 75 horsepower, enough to push the SS1 to over 80 mph. Additional improvements included adding a water pump, to circulate coolant more positively than was possible with the old thermo-syphon system, and sweeping the exhaust manifold forward on its way down from the engine, to keep heat from the cabin.

That engine work was carried out by Standard, but clearly Lyons was serious about making a good car. His SS1 was a definite success. During five years of production, 1932 through 1936, sales of all three series and all four body styles totaled more than 4200.

It was one of the third-series 1934 chassis, fitted with the 2.7 engine, that William Walmsley chose as the basis for his final contribution to the company he was shortly to leave. This was an open four-seat Tourer shortened in both body and chassis to become a "proper" two-seater. Little information seems to have survived about this one-off project, but in appearance it must have strongly foreshadowed the later SS 90/SS 100. The greatest known difference is that Walmsley's sports car carried its spare tire in a lay-down position in a notably long tail.

That car was apparently finished in March 1934. So were Lyons and Walmsley, just about, but the one William evidently saw merit in the other's toy. Late that

*The second-series SS1* (both pages) *had an airier cabin with more rear-seat room. Stylish appointments inside and out ensured that already-famous high-value Lyons touch.*

year, even as the 12-year partnership was being dissolved, a second two-seater was built. Finished in December, this is the car regarded as the definitive prototype of the SS 90.

As would the production cars to follow, the prototype SS 90 rode a regular SS1 chassis with 15 inches cut from its middle. The X-member was removed at the same time, shortened a little, then welded back in with the remaining structure. Atop this modified frame was mounted a body that resembled the Tourer's but which was, in fact, a new piece of work and lurked lower between tall "racing" wire wheels. The radiator shell and the huge, multi-louvered engine hood were both squarer and seemed to bulge with unspeakable power. Giant Lucas headlamps stood on great, manly chromed braces. In delightful contrast, four of the most delicate fenders ever drawn seemed more mist than metal.

Unique to this prototype SS 90 was the tail treatment. A truncated cone of body metal that concealed the gas tank, it finished in a cutout carrying the spare wheel at a steep forward cant. Apparently to open some space for luggage and a folding top, all succeeding SS 90s had more rearward, fully exposed "slab" tanks and vertically mounted spares. What was thereby lost to aesthetics was regained up front, where the side louvers and hood rear edge, initially vertical, were slanted back to match the forward edge of the door and the windshield. The "coach-built" body construction itself was conventional for the time, a framing of well-seasoned ash supporting aluminum panels. The fenders, or "wings" in the particularly apt British term, were aluminum as well.

The new wheelbase was 104 inches, which reduced the much longer SS1's ocean-linerish turning circle of 38 feet to a still not very nimble $35\frac{1}{2}$. The track measurements, front and rear, remained at 54 inches. Cus-

*A quartet of side windows lent the SS1 its Four-light Saloon tag. For this '34 model, Lyons kept the base engine at 2.1 liters but the larger six grew from 2.6 to 2.7 and delivered about 75 horsepower for an 80-mph top speed.*

tomers had a choice of tire sizes, either 5.25 or 5.5 Dunlop "90"s mounted on big 18-inch rims. Conventional in most respects for the time, the beam axle/leaf spring suspension did offer one feature of special interest to anyone who supposes new ground was broken some 50 years later: these 1935 shock absorbers, though non-hydraulic, were Andre Telecontrols, adjustable from the cockpit.

Only the larger Standard engine was offered in the SS 90, the "2½-Litre" that actually displaced nearer 2.7. Compared to its tune in the SS1, it had a higher compression ratio—all the way out to 7.0:1—and a sportier camshaft profile, two carburetors instead of one, and a new cast-aluminum oil sump with greater capacity. The last was significant, because this engine still had a reputation for marginal cooling, and its connecting rods were still Dural aluminum alloy. Elevated running temperatures would open the rod clearances enough to lower oil pressure. By going to the trouble of improving the oiling system, SS Cars was tacitly endorsing the use of its new sports car as a sports car.

In point of fact, William Lyons himself endorsed the idea overtly at least once. Bringing one of the early 90s to show off at an SS owners' club meet in his old hometown of Blackpool, he proceeded to set best time of the day in a driving test. So this definitely was the fastest Lyons car yet.

In absolute terms, though, the SS 90 wasn't very fast. No independent tests seem to have been carried out, so any performance numbers are anecdotal, but at best the engine put out no more than 80 bhp, so the car was capable of only approaching a genuine 90 mph. Brisk enough, probably, given the roads, tires, suspensions and brakes of the day. But it fell short of generating the aura William Lyons by now had in mind.

The car went on sale in March 1935, but apparently the factory was already working on its successor. Only 22 production SS 90s were turned out before the SS Jaguar 100 was ready that fall.

Engine improvements were the big news. Thanks to work done by consultant Harry Weslake for the new SS Jaguar sedan, introduced at the same time, the SS 100 arrived with an overhead-valve cylinder head. The twin carburetors, now SUs instead of the older RAGs, were still on the left side, but the exhaust manifold had been moved to the right to make an efficient cross-flow design. Even for the day, this was not startling technology, for the head was cast in iron and operated the valves through pushrods. But Weslake's genius had somehow conjured up startling results. Lyons had asked him to raise the output of the side-valve 2.7-liter from 75 horsepower to about 90. With his ohv head, Weslake raised it beyond 100; on the dynamometer, the test engine showed 105 bhp at 4500 rpm. The production unit was almost as good: 104 at 4600.

Though little changed from the 90 in basic dimensions and basic chassis design, the 100 arrived with a somewhat different steering mechanism and much bigger brakes operated by rod, rather than cable. Suspension was revised, too, with some parts adopted from the new sedan. Curiously, front shock absorbers were now a combination of Hartford friction and Luvax hydraulic shocks, but only the latter were used at the rear. Tires were the same size, but the wire wheels were now supplied by Dunlop instead of Rudge-Whitworth, and had different hubs.

Visually, the biggest alteration from 90 to 100 occurred at the rear, where the gas tank's back wall now slanted forward, and thus so did the spare wheel. However, someone standing toward the nose of the car could easily notice a difference in the front wings, where the leading edges had a more florid droop. There were also some bumper, grille and, of course, badging differences.

The SS Jaguar 100 went on sale in March 1936, and was an immediate hit. It was a faster car, also a better car, and it listed for a still-remarkably low £395. The

*A harbinger of the sports cars to come was the SS1 Tourer* (both pages). *With its side-valve six of up to 2.7 liters* (top), *the Open Four-Seater Sports was the first Swallow product to be used by its owners for serious work in rallies and gymkhana-like events.*

sporting public responded. Over the next two years, well over 100 "100"s were sold, and many did well in various sorts of speed and endurance events, including international rallies.

According to a contemporary road test in *The Motor*, Britain's second-oldest automotive journal (after *The Autocar*), a 2.7-liter SS 100 could do 0-60 mph in 12.8 seconds. At the time, that was quick enough to shut down most locally made opposition. Similarly, the quarter-mile time of 18.6 seconds was considered "outstanding." As for top speed, the magazine timed its car at 96 mph. Another publication got a little less, but even with the full-sized windshield in its upright position the car could hit 90. In normal motoring, 75 was regarded as an easy cruising speed.

More speed came for 1938. In conjunction with developing its new "all-steel" sedan, SS Jaguar bored and stroked the sturdy old Standard six-cylinder to 82 x 110 mm. This brought its capacity up to 3485.5 cc (212.7 cid), a resounding 31 percent more and nearly 75 percent over the old 2.1-liter with the same basic block. To handle the extra stress, the chrome-iron crankcase was made stiffer; to handle the extra gas flow, the ohv head was revised to have six individual exhaust ports. Steel connecting rods were adopted, finally, to cure the old hot-oil pressure problem. On a compression ratio of 7.2:1, horsepower was said to be 125 at 4500. Inevitably, price went up too, but only to a still-very-competitive £445.

It was well worth it. Genuinely much quicker than the 2.7-liter model, the "3½-Litre" SS 100 amazed the enthusiast press of the day. Zero-to-60 mph acceleration times dropped below 11 seconds (to 10.4, said *The Motor*) and, at least with the main windshield folded down flat, the "100" proved to be a 100-mph car by one or two mph. These figures may not look very impressive today, but to better them in Great Britain during the late 1930s, one had to buy a much more expensive car—or maybe a 998-cc Vincent-HRD motorcycle.

Actually, driving the SS 100 would have had points of similarity to riding a big bike. Although a fold-

*Like all SS1s, the Tourer impressed with its high level of fit and finish at a cost of no more than £340. The 1936 model pictured had Andre Telecontrol shock absorbers, which could be adjusted with the flick of a switch* (above), *easing the convertible's passage over the often-poor European road surfaces of the day.*

ing top (or "hood") was provided, and while this allegedly could be set up inside 60 seconds, Britons who ran sports cars in those days took pride in taking as much of their motoring *al fresco* as was humanly endurable. In this and other ways, this first great Jaguar—considered quite modern at the time of its introduction—now seems to hark back to very early times indeed.

It's easy to climb aboard an SS 100. The rear-hinged door opens very wide, and the pillowy leather seat is not slung especially low. However, the four-spoke steering wheel is both big—18 inches is the standard diameter—and rather close-in to the belt buckle. Happily, it's easily adjustable for distance and, with tools, for rake angle. Even when closed, the door doesn't seem to be there because of its very deep elbow cutout. There's not much but air all around, in fact, and you feel more atop the car than within. But your legs have vanished into a long, narrow tunnel and this creates a compensating feeling of security, as if you've been holstered.

That wondrous great bonnet with its four rows of louvers dominates the view. It seems remarkably long; it actually measures 49 1/8 inches along the chromed piano hinge linking its two "butterfly" sections. Jutting up proudly either side are the huge headlamps, borne by the flaring wings.

Not far above your lap is a curvaceous double-lobe instrument panel displaying a sextet of elegantly white-faced gauges, including a 100-mph speedometer down low to the right of the steering column, and a 5000-rpm tach to the left. The latter, which works counter-clockwise, has a red sector (lacking on some cars) beginning just above 4500. Most secondary controls are spread across the rest of the panel, save two levers on the steering-wheel hub for lighting and ignition timing.

The SS 100 still had the longish 104-inch wheelbase and substantial 53-inch track, but it was only 150 inches long overall and 63 wide, so it didn't take up that much road space. It was fairly light, too. A figure of "23 hundredweight" is usually given for a typically equipped car. A "hundredweight," often expressed as "cwt.," is one

*Jaguar's first sports car was the SS 90. It bowed in March, 1935 and was visually and dimensionally very similar to the SS 100 pictured on both pages. Only 22 SS 90s were built before the SS 100 debuted a few months later, bringing style, speed, and success in competition* (below) *that wrote a new chapter in the Jaguar saga.*

*The authority of its prow, the classic simplicity of its two-seat cockpit, the enduring dips and swoops of its fenders—say "sports car" and a generation of drivers responded, "SS 100." Perhaps best of all, it was obtainable, hitting the market in 1936 at a remarkably low £395.*

of those quaint measures still dear to the English, like "stone" (used to express a person's weight in 14-lb chunks) or "Guinea" (a coin of the realm equal to 1 £ plus 1 shilling). A "cwt." is 112 pounds, so a road-ready SS 100 apparently tipped the scales at 2575 pounds. A mere bagatelle to the mighty 3.5-liter, 125-horsepower Jaguar engine.

What fun it must have been to set off in a brand-new 100 for a bracing canter along the narrow, twisting, usually quiet country roads of its homeland. The ohv engine retains much of the steam-engine torque delivery of the earlier side-valve, but there is more of it, so the six big aluminum pistons pull smoothly and eagerly from as low as 500 rpm. The four-speed transmission uses a conventional H-pattern with reverse left-forward (and "LIFT," as marked on the knob). Its movement is short and not too heavy. Luckily, a tendency in early gearboxes to break teeth had been cured. However, first gear lacks synchromesh, and while the upper three have it, they still require a deliberate, sensitive hand.

With only 2.3 turns lock-to-lock, the steering is pleasantly responsive, yet light enough to make the car feel easy to handle. The front-end geometry—so long proven on Standards—is such that shimmying, tramping or other bad beam-axle habits seldom are noticed. The rear axle, however, prefers England's typically smooth roads, as poorer ones can set up a wheel hop. Coupled with the car's natural tendency toward oversteer (tail-sliding) at the limit, turns can abruptly turn wild. Words such as "unpredictable," even "skittish," still aptly describe the cornering behavior. The ride on rough surfaces is pretty harsh, too, but drivers of this period believe discomfort is the necessary price of responsive handling.

Driven ruthlessly, the SS 100 can use up its tires rapidly, and an eye has to be kept on the water-temperature gauge. But the braking power with the bigger drums is considered entirely adequate for the performance and driving techniques of the day. And over long, arduous routes, often against time constraints, the car has begun to build up a surprising record of sturdy reliability.

*It was the first great Jaguar, and it was very fast for its day. With the original 104-bhp 2.7-liter ohv six, the SS 100 ran 0-60 mph in 12.8 seconds and topped 95 mph. An increase to 3.5 liters for '38 brought 125 bhp, cut 0-60 times to under 11 seconds, and jacked top end to just over 100. Hard drivers felt the nose grow light near maximum speed and the stiffly sprung tail get skittish in bumpy turns.*

The SS 100's most unsettling characteristic is the impression it sometimes gives at high speeds that the front end is flying. Whether this is caused by the rakish "wings," or perhaps by weakness in the steering mechanism, will never be resolved. But some drivers who wind the speedometer right to the 100 mark—or who encounter strong wind gusts at lower velocities—feel the steering go light and vague.

Its undeniable quirks, coupled with a widespread first impression that the extravagant-looking sportster was just too pretty to be taken seriously, held some critics back from immediate acclamation. But many of those who took time to really get to know the SS 100 found that it grew on them. As *The Autocar* put it: "An outstanding feature recalled of the SS was that, whilst one felt sufficiently at home in it from the beginning, there came a stage, after perhaps a couple of hundred miles, where one suddenly found a great deal more in the car than there had seemed to be at first—not so much in sheer performance as in confidence in it."

This passage is quoted from *Jaguar Sports Cars*, a book by marque expert Paul Skilleter, who went on in his own words to make this important point: "Thus William Lyons had managed, with what was really his first production sports car, to achieve that 'thoroughbred' feel which was to remain present in all successive Jaguar sports cars; a certain quality by which a personal relationship can be established between car and driver."

It is just this quality that forgives a multitude of niggling little sins. To the driver who deeply enjoys driving, a car shows a personality, and when it is a willing and eager car, a car that seems to have been born for the joy of the open road, such a driver dismisses foibles and imperfections. They're irrelevant to the relationship.

That was the root of William Lyons' success as an automaker. His cars would always be flawed, but they would never be soul-less transportation appliances. They could be loved.

A telling vignette about Lyons' own involvement with driving occurred one sunny spring day in May of 1938. At Donington Park, then home of the British Grand Prix, he joined his chief engineer, William Heynes, a third

*Rear-hinged doors, a steering wheel 18 inches in diameter, elegant, white-faced gauges, the short-throw four-speed gearbox, delicate dual windscreens, steam-engine torque—all essential elements of the SS 100's character.*

SS 100 driver, and one in an SS 90 for a little demonstration race. So anxious was the SS Jaguar chief to get going that he twice jumped the start and had to be called back. Once fairly away, he drove, according to a contemporary account, "with the most awe-inspiring determination," and won.

Other 100s in other hands were winning other events. Not great events; SS Jaguar's sports car was not yet a race car. It was never meant to be. Yet it had that appealing "let's try!" temperament, and so people did try.

Perhaps the most venerated of the SS 100s in competition was one owned and modified by the factory itself. Known to Jaguar enthusiasts as "Old No. 8," from its chassis number 18008, it began as the 2.7-liter street car that won the top prize in the 1936 International Alpine Trial, a grueling long-distance Continental road rally. Bit by bit, over the next several years it was hopped up and stripped down for track work. It lost its fenders for better aerodynamics, but gained a 3.5 engine. For better weight distribution, its engine was moved back in the chassis and both its axles were moved forward. In 1937 it won a race at the old Brooklands banked speedway at 118 mph. Two years later it lapped the same track at 125. Running on a 14.0:1 compression ratio and exotic fuels, it showed 169 bhp at 4250 rpm on the test bed. Development resumed after World War II, and in 1947 the engine was made to put out 171 bhp at 4500 rpm. Even then people couldn't keep their hands off the grand old thing, and a later owner reported he'd turned it to 4800 rpm on a chassis dyno, at which rate the reading was 172 bhp at the rear wheels. He reckoned that meant about 215 at the flywheel. Old No. 8 was then over 30 years of age.

Of course, the factory had long since turned to other things, having ended SS 100 production in 1939. The number built is believed to be 308. Adding prototype and production SS 90s brings that up to only 332.

Small numbers, but big impact. The SS sports cars had been seminal in design and performance. They also were smashing commercial successes for a small company striving to make its mark. With them was struck the pattern for every sporting Jaguar to come—the image and personality that still fire enthusiast hearts today.

CHAPTER VI

# XK 120: ASTONISHING AFTERTHOUGHT

When the English want to convey great excitement, they invoke the image of a "cat amongst the pigeons." What more appropriate occasion to employ this favored old phrase than the sudden, surprise presentation at London's 1948 Earls Court Motor Show of the new "Super Sports" Jaguar?

A low-slung, gleaming, metallic-bronze roadster, it must have been a truly startling apparition there on the spotlit floor amid a flock of upright black and grey "saloons." Exotic beyond belief for that somber time and place, the XK 120 offered not only spectacularly streamlined styling but a magnificent new double-overhead camshaft, 160-brake-horsepower engine. "I can put two miles into the minute," was the proud proclamation of the XK's model number. "Why, that's racing-car performance!" was the reaction of the stunned multitude.

*The car that triggered a tectonic shift in the fortunes of Jaguar: the XK 120. William Lyons said it went from conception to full-size mockup in less than two weeks. Perhaps that explains the simplicity and perfection of its line.*

Only the rarest of automobiles have ever had such impact on their introduction. But the mesmerizing new Jaguar was more than just a gorgeous new car; it was a symbol, and a statement. As a sports car, it was made in a spirit of buoyant celebration. It represented nothing less than a complete release from the gloomy fears of the war years. It exalted the joy of living free after nearly a decade of threatened servitude. More: The acclaim (and orders!) showering in from all continents proved to Britons that their battered, blackened nation could still produce something of international value.

The XK 120 went on to prove itself a sound machine in the honest crucible of competition, winning innumerable races and arduous rallies, and serving as the launching platform for several great competition driving careers. Its engine also achieved some of the proudest moments of British motorsports history, powering Jaguar racing cars to five outright victories in the punishing 24 Hours of Le Mans, thus establishing the marque on a plane of honor with the hallowed Bentley.

On the road, the XK 120 stood perhaps not above, but certainly apart from the other sports cars of its time. It was so fast and yet so comfortable, so beautiful, so fresh, so amazingly ambitious, as to be innovative in a curious way: It was not trend-setting. It was too special to be copied. It was its own trend.

So different was the design, so strong the character, that there were some who reacted negatively. In fact, it was impossible to be indifferent to the Jaguar. There was no middle ground; one either had distaste for it, or wanted it desperately. But for those to whom there was a strong appeal, it was an appeal that ran deeper than mere automotive values. The XK 120 was liked and desired by people who loved cars, and also by people who didn't necessarily care a fig about cars.

As with great works of art, this car has not dated. In detail, yes, but its basic lines and proportions and its sheer audacious aesthetic thrill are still hauntingly lovely more than 40 years after it first sprang into the spotlights.

Also still true is the fact that the XK 120 is the most significant car Jaguar has ever made. Its predecessor, the SS 100, had made the company important. The 120 made it immortal.

And the oddest thing about it is, it came about almost by accident. In William Lyons' own word, the 120 was an afterthought.

*Chief engineer William Heynes oversaw development of the magnificent XK engine. Unveiled in 1948, the twin-cam six was Jaguar's soul into the 1980s.*

Lyons had started out in the motorcycle-sidecar business, and his first automobile body was a two-seater, but he well understood that his company's fortunes were based on multi-passenger sedans. So were his SS sports cars of the 1930s. Their bodies were specially designed, but almost everything underneath—engines, drivelines, suspension, etc.—was simply borrowed from the contemporary high-end SS sedans. In a sense, then, this former "packager" was now repackaging his own product. No harm in that. Baking sports-car cakes from bread-and-butter passenger-car ingredients was sound policy both economically and technically, and many other automakers have followed the recipe. Lyons himself would do so again.

Work toward the new Jaguar sedan that would sire the XK 120 sports car actually began before World War II. Having created an all-new four-door "saloon" with an all-new name, SS Jaguar, in 1935, and having shifted its body from wood-framed to all-steel construction in 1937, Lyons decided the next step was a more modern chassis. Thus his chief engineer, William Heynes, began in 1938 to investigate a variety of independent front suspension designs. At least one such car was built and running before war broke out in the fall of 1939. During the dark years that followed, Heynes and his engineers continued to gain experience with independent suspension fitted to experimental military vehicles.

When the renamed Jaguar Cars, Ltd. was finally able to restart its passenger-car production lines late in 1945, it simply issued slightly restyled versions of its prewar sedans. And why not? There was little need for anything else. As in America and other countries, a public deprived of consumer goods by the long war cared far less about new automobile designs than just having new

*Jaguar built its two-seat sports car on a shortened sedan chassis using many sedan-based components. Its 102-inch wheelbase was two inches shorter than the SS 100's, but the XK was nearly two feet longer overall. Early cars had aluminum body panels over handmade ash framing. Demand soon dictated that the body—less hood, trunklid and doors—be redesigned for mass production in steel. Rear-fender spats were an XK 120 signature. This early study shows Jaguar's attempt to fit them in conjunction with wire wheels, which had more prominent hubs than the steel items.*

automobiles, so these Jaguars sold relatively well. But everyone realized that the postwar seller's market wouldn't last forever; sooner or later, all-new models would be needed to replace the prewar holdovers. In Jaguar's case, that meant a more-modern new sedan. Fortunately, it was on the way.

Apparently because it employed the fifth iteration of Heynes' independent front-suspension designs, this car became known within the company as the "mark five." The designation survived to 1948, when the model was introduced as Mark V, even though no previous Jaguar had been given a mark designation. (As a point of argument, one could count the entire series of SS1s from 1931 to 1936 as William Lyons' first "mark," in which case the 1935 SS Jaguar would be the second and the 1937 all-steel version of that might qualify as the third. The interim model introduced in 1945 could thus be considered the Mark IV, as indeed it has sometimes been called in retrospect, especially in America. But mark designations under "five" were never official before this.)

No matter. The Mark V was a timely advance on Lyons' sedan concept. For example, its new independent front suspension incorporated torsion bars as the springing medium, something like that of Dr. Ferdinand Porsche's *Volkswagen* and the forthcoming German sports cars bearing his name. The Mark V also featured modern hydraulic brake actuation and a particularly strong, deep-sided chassis.

There was one other important technical advance under development, though it was not intended for the Mark V. It was a new engine, Jaguar's first totally in-house powerplant, being readied for yet another new sedan scheduled to debut in 1950 as the Mark VII. The fruit of forward-planning sessions between Lyons and his engineers during the war years, it was still an inline six displacing about 3.5 liters. But instead of the pushrod-operated overhead valves of the old Standard powerplant, it adopted from racing practice the exotic concept of double overhead camshafts operating inclined valves in hemispherical combustion chambers. But this so-called XK engine was not primarily intended for racing; rather, it was engineered to be powerful, efficient, smooth, quiet, and durable in everyday passenger-car service.

That simple description belies the enormous amount of thought, time and engineering talent that went into the XK. It was no overnight success.

Working under overall direction from Heynes, the new engine's detail-design team consisted primarily of Walter ("Wally") Hassan and Claude Baily, plus Harry Weslake as an outside consultant. Actually beginning before the end of the war, they concentrated most of their development on a series of four-cylinder engines.

They started with fours because, based on Lyons' market experience, he anticipated manufacturing a two-tier range of postwar sedans. A comparatively small car powered by a modest four in the under-2.0-liter class would form the solid commercial basis. More well-to-do customers could choose a lower-volume premier model with a six-cylinder of about 2.5 liters. Looking farther into the future, Lyons then wanted to be able to offer additional model levels, some powered by eight- and even twelve-cylinder engines. For economy of production, all of the later powerplants were to be derived directly from the basic four.

"Basic" is a misleading word, however, because Lyons was determined that all of these engines would enhance Jaguar's reputation for offering power and beauty at realistic prices. As his body designs had stunned the world, so now must his powertrains.

Heynes and his men began wrestling with the novel problem of producing an inherently exotic engine economically. They wrestled only on paper at first, drawing up a series of hemi-head fours, each carrying the designation "X," for "experimental"; XA led to XB, and so on.

*Astute promoter William Lyons arranged for the press to witness the XK 120 live up to its name on a closed section of Belgian highway near Jabbeke in May 1949. Virtually stock except for a streamlined belly pan, smaller windscreen, and cockpit tonneau, it blew by the journalists at better than 133 mph. With the stock windshield and convertible top up, it averaged over 126. Sensational!*

They weren't ready to try it in the metal until they got to XF, their sixth design. This displaced a mere 1360 cc (83 cubic inches) but thanks to efficient dual overhead camshafts, it put out a lot of power. Too much power, actually. Its block wasn't strong enough.

The next step was to fall back on the well-proven four-cylinder, 1.7-liter Standard block that SS Jaguar had been using for years. Atop this was tested the XG head, another design entirely, one that had its inclined valves operated by pushrods. It was a layout very similar to that of the six-cylinder BMW 328 which Heynes, thanks to the generosity of a friend, had been able to study closely. Heynes' cheek extended to road-testing this provisional powerplant in a BMW sedan—and to this verdict: The engine's design, he said, was inherently too rough and noisy for a Jaguar.

When the team got to the XJ, they were close to what they wanted. This engine was back to the twin-cam configuration and, in 2.0-liter form, it worked so well that they went ahead with a six-cylinder, 3.2-liter version. That was even better—smooth and powerful. The XJ-six's only deficiency was its torque output, so the stroke was increased to give 3.4 liters. The design designation was advanced to XK, and the job was done.

By this time, obviously, Lyons had abandoned his original displacement ceiling of 2.5 liters and he was reexamining his two-tier model plan, as well. Fears of postwar constraints involving taxation and petroleum hadn't materialized, at least not to the point that people were refusing to buy larger cars. Besides, the prospects in overseas markets, particularly the United States, were turning out to be far richer than anyone had dreamed.

The idea of smaller, four-cylinder Jaguars did remain alive for several years. Several dozen XK-fours were actually built but tucked away in a corner; none ever reached the production line. In 1948, as in succeeding years, it seemed far more sensible to concentrate on the bigger, more profitable models.

However, the stillborn four did make one tangible contribution to Jaguar's performance history. Just before

*Perhaps the most celebrated XK 120 was Ian Appleyard's roadster, here still wearing its Alpine rally shield. This is an early aluminum-bodied roadster, as identified by the rubber grommet at the base of the windscreen. Some speeding drivers were surprised to find the engine cover spring loose and fly back over their heads. The leather strap across the hood prevented such occurrences.*

the XK 120 would be unveiled, arrangements were made with well-known land-speed-record driver A.T.G. "Goldie" Gardner to install a four-banger in his streamliner, in which he normally used an MG engine. Called the EX-135, the tiny missile was taken to a straight superhighway near Jabbeke in Belgium. The experimental Jaguar engine, with a displacement of 1996 cc and race-tuned to give 146 horsepower at 6100 rpm, pushed Gardner to a two-way average speed of 176.694 mph. It was a new world record for the 2.0-liter class, and the publicity was duly exploited by Jaguar to launch its new "Super Sports" at Earls Court a few weeks later.

In fact, early XK 120 sales literature spoke of a sister model, an "XK 100," that would be identical except for a twin-cam, four-cylinder engine. Some observers have wondered aloud just how serious Lyons was about producing such a car. He never would, but their cynicism was unable to stem the praise showering in from all quarters for the six-cylinder model.

Jaguar's six-cylinder XK—strong, dependable and versatile as well as very handsome—would go on to establish itself one of the most successful automotive powerplants of the century. In its initial form, it had a bore of 83 mm and a stroke of 106 (3.27 inches x 4.17) for a displacement of 3441.14 cc (209.99 cubic inches). That stroke dimension happened to be the same as in the old pushrod "three-and-a-half" Standard-based engine, but that was about the extent of the family resemblance. The only other respect in which Jaguar's new engine built on experience with the old was its very strong iron crankcase, which mounted the massive six-throw crankshaft rigidly in seven main bearings.

The rest of the engine was designed according to the latest piston-engine science. Inside the aluminum alloy head, under beautifully polished covers, a pair of chain-driven camshafts operated two rows of valves each

*The inspired XK six in its original 3.4-liter form* (right) *made 160 bhp at 5000 rpm. It powered Jaguars to checkered flags in numerous contests, including Ian Appleyard's '50 Alpine Rally win* (opposite). *Coventry never put the four-cylinder variant into production. But an experimental 146-bhp version took "Goldie" Gardner's EX-135* (below) *to a 2.0-liter record 176.6 mph average in 1948.*

inclined at an angle of 35 degrees to the cylinder axis. The advantage of the overhead cams, of course, was to reduce valvetrain weight to a minimum, thus allowing high rpm with minimum noise and wear. By inclining those valves in hemispherical combustion chambers, Jaguar achieved the maximum possible valve diameter for the best breathing.

Other important details included Weslake's careful shaping of the intake ports, which ensured a degree of turbulence to promote thorough burning of the mixture. To achieve the standard initial compression ratio of 8.0:1, the cast-aluminum piston crowns had pronounced domes fitting up into the hemispherical combustion chambers. Carburetion was by a pair of one-and-three-quarter-inch SUs mounted on the right-hand side. The exhaust was on the opposite side, its manifolding a brace of three-branch castings finely finished in black enamel. Quoted horsepower was 160 at 5000 rpm, with a torque output of 195 pounds/feet at 2500.

Although the XK was designed for primary duty in sedans, it certainly did look like a racing powerplant. Of course, it was designed by people who enjoyed racing and who had done quite a bit of it—and who knew that the perhaps-surprising competition successes of the SS-series of cars had done wonders for their company's reputation. Right from the beginning, it must have been in their minds to try the XK on the track at some stage. After all, the beautiful new twin-cam had a lot more latent performance built into it than was strictly necessary for a passenger car.

Introducing the XK by building a sports car around it must have been equally tempting. Consider the advantages: A sports car would make a more appropriate first home for the new jewel-like engine than a mere sedan; it would attract more attention from people who talk about—and write about—interesting new cars; it would offer a chance to gain production and in-service experience with the all-new powerplant in limited numbers before committing to large-scale production; and it would keep faith with the cadre of SS 100 loyalists who were still racing and rallying the prewar sports car, and often winning with it. If a clearly obsolescent old sports car were still so competitive, imagine the good a modern one could do.

Thus the deed was put in hand. On a summer's day in 1948, Lyons ordered that an 18-inch section should be cut from the middle of a prototype Mark V chassis. Then he called together his favorite "panel beaters" and had them start on a full-size metal-on-wood mockup of a racy, open-cockpit two-seater for the shortened chassis.

In later years, Sir William recalled that the entire job took less than two weeks. That's an astonishingly short time for a process to which he normally devoted months, and in some cases a year or more. It helps explain the marvelous, elegant simplicity of the XK 120's lines; deft, sure strokes of a master who knew exactly what he wanted from the start, and who forbade himself second thoughts.

*Chief rival of Jaguar's SS 100 in the '30s was the BMW 328* (top, both pages). *Heynes road tested the German car and studied its hemi-head engine.*

*The BMW 328's influence on the design of the subsequent XK 120* (bottom, both pages) *is clearly evident, particularly in the front-end treatment.*

Actually, the design had been evolving in his mind, and even in the metal, for a lot longer than two weeks. As far back as 1938, 10 long, tumultuous years earlier, Lyons had produced a semi-streamlined body on the SS 100 chassis. It retained that very traditional car's radiator shell and engine hood, and also the big separate headlights. But in a style that had begun appearing on some Italian, French, and American cars, the fenders enclosed the wheels in long, tapering shells like those on aircraft landing gear. The rear wheels were covered by sheetmetal spats, and the cockpit by a cozy, rounded coupe top. This model was given a list price of £595, some £150 over the normal 3½-liter SS 100, but it never went into production. Reportedly, the interior was more stylish than habitable.

Certainly that prior experience went into Lyons' conceptualization of his 1948 sports car, but there can be no question that an even more important influence was the BMW 328 of 1936. Along with several other *avant-garde* models of the decade, the body design of this German two-seater attempted to unify and round-off the separate and angular elements of hood, fenders, and headlights without losing their traditional identities. Mechanically, the 328's notable points were a strong tubular frame, an independent front suspension, and an effi-

*Zero-60 mph in 10 seconds flat and the quarter-mile in 17 was "acceleration such as most drivers have never even imagined,"* The Motor *said after its run with an early XK 120. Other testers got slower times, but the possibility of such performance for only around $3000 U.S. meant the new Jaguar* mattered.

cient six-cylinder engine with hemispherical combustion chambers and inclined valves.

All this made for a modern-looking, comfortable, fast car with good handling and roadholding. Though pricey, the 328 became very popular in prewar England, and proved to be the SS 100's most direct rival on both road and track. William Lyons certainly knew all about it, and, as we've seen, so did William Heynes. Thus in several important respects, including engine, chassis, suspension, and body, the new postwar Jaguar sports car, while not a copy of the prewar BMW, was obviously a child of it.

The first known tangible expression of Lyons' evolving postwar thought was a styling mockup that he may have had hammered together as early as 1946. It was not as large as would be the 1948 "Super Sports" model, being planned for a small four-cylinder engine, but the general shape was predictive. In front of an invitingly open, low-sided cockpit not unlike those of older aircraft—or of the early Swallow sidecars—extended a long Roman nose of a hood that recalled the bold snouts of Battle-of-Britain fighter planes. This hood carried the tall, ovoid grille opening and, from appearances, was hinged at the back so it opened like an alligator's upper jaw (already a feature of Detroit cars for some years). Front and rear fenders were blended together with the doors in one long, wasp-waisted swoop, and spats ("fender skirts" to us Yanks) covered the rear-wheel openings.

So far, that description would fit the first XK 120 just as well. The main difference was the position of the headlights, which in this early styling exercise were very low down in the fronts of the rounded fenders. On the XK 120, Lyons moved them back up to their time-honored—and arguably more practical—places either side of the grille.

What is much harder to describe is the utterly different look of the XK 120. Though actually a simpler shape, its body was finished, was a coherent whole, made a complete statement in a way few cars ever have. It was impossible to improve on it—as would be sadly demonstrated by a variety of later attempts both within the factory and without. The body seemed not so much to have

*Steel disc wheels and rear-fender spats accentuated the XK 120's gracefully streamlined lines. Data is sketchy, but an aluminum roadster in street trim probably weighed between 2850 and 2950 pounds. Steel versions like the one pictured bowed in 1950 and were not significantly heavier.*

been manufactured as to have grown to fulfill some ancient genetic mandate. It may be fanciful to suppose it was deliberate on William Lyons' part, but it is possible to see in this exquisite Jaguar an actual feline image. From a viewpoint either side of the front wheels, the repeating, forward-curling curves of fenders, hood and headlight nacelles suggest the paw of a giant cat.

The great cat leapt onto the world stage on October 27, 1948. It likely took only a few hours for Jaguar management to realize that things were completely out of hand. Everyone, it seemed, had heard of the car, had come to Earls Court to see it, and wanted to go home with it. Immediately.

And the clamor didn't stem solely from the XK 120's dramatic appearance, for everyone who cared anything about engines couldn't wait to put their foot into that splendid-looking twin-cam. To the ordinary sports-car enthusiast, such an advanced design had long been daydream material, the stuff of Grand Prix racing cars and a very, very few of the most fabulous, impossibly costly road machines—Alfas, Bugattis, Duesenbergs. Yet here was little Jaguar Cars, Ltd. of far-off Coventry in middle England poised to produce such a thing for the masses. The price? Had to be a misprint: The XK 120 was going to sell, pre-tax, for a mere £998. Call it 3000 contemporary U.S. dollars.

That meant this car *mattered*. By being so accessible to so many, this one, startlingly advanced Jaguar was going to have a major effect on the entire world of sports motoring. And, indeed, its departures from established sports-car traditions and philosophies, at least as then understood in Great Britain, were going to take some coming to terms with.

Partly, it was a problem of pragmatism. The new Jaguar was simply so practical, so sensible. And so comfortable! Yes, it had a body fashioned of lightweight, fragile "aluminium." But it was streamlined. All those voluptuous curves; how was one to see the wheels for cornering? To see the corners for parking? Why, inside its shapely tail there even lurked a proper luggage boot. What was this modern world coming to?

Seriously, though, the XK 120 does seem to have

forced many who considered themselves enthusiasts of the traditional "sports car" to examine just exactly what they meant by that term. The new Jaguar was a very fast open-cockpit two-seater, but it was also larger, heavier, and something less than nimble next to most of the tear-abouts then popular.

True, the XK 120's stance on the road was a bit more compact than that of its SS 100 predecessor. At 102 inches, the 120's wheelbase was shorter by two inches and with track measurements of 51 inches front, 50 rear (versus 54/54), it was narrower, too. But at 174 inches overall, the newer car's enveloping body was nearly two feet longer. It was heavier, too, by several hundredweight.

The extra weight was a consequence of Jaguar's frugality in the borrowing of so many components from its new Mark V and upcoming Mark VII sedans. The massive two-rail, box-section chassis was virtually unaltered, save a little shortening. Ditto suspension: independent fore via torsion bars, telescopic shocks and outer ball joints; non-independent aft via leaf-sprung live axle and lever-type shocks. It was all designed to support something far more substantial than a two-seat body. Likewise the recirculating-ball steering, the hydraulically operated four-wheel drum brakes, the steel disc wheels, the single-plate clutch, the four-speed manual transmission (still synchronized only on the top three) all made for inexpensive production and long, heavy duty. Even that marvelous-looking 160-horsepower engine was essentially an iron-block passenger-car powerplant designed first of all for utter reliability in sustained use under the worst of conditions.

None of these plebeian practicalities mattered a jot to the throngs of desperate would-be XK 120 buyers jostling and elbowing each other for a glimpse of the sleek bronze prototype roadster. The original plan had been to assemble about 200 cars. Demand, clamorous and relent-

*Gutsy Jaguar wasn't shy about running its XK 120 in such storied events as the 24 Hours of Le Mans, the Mille Miglia, and the Targa Florio. Designed purely for the road, it was in a bit over its head against the hard-core sports/racing machines, but the experience was invaluable and the publicity did much to cement its reputation as a premier production automobile.*

less, indicated that the potential market was in the thousands. But that was impossible. The XK 120's heavily labor-intensive body construction technique, whereby stamped aluminum panels were mounted on handmade ash framing, was completely inadequate for such production volumes.

Accordingly, the entire body (less hood, trunklid and doors) was redesigned for mass production in steel. Meantime, Jaguar pressed ahead with handmade aluminum cars to satiate as much market hunger as possible. And to take advantage of the intense interest on the part of the enthusiast public, it fielded the speedy new sports car in competitions it had a very good chance of winning.

First order of business was to convince the doubters that the XK 120 really could do 120 mph. In the spring of 1949, after some clandestine early-morning runs outside Coventry, a car was taken across the Channel to a straight section of Belgian *autoroute* just inland from Ostend, near the village of Jabbeke. Again, a careful preliminary test proved the potential. For the public runs on Monday, May 30, local police closed off half of the divided highway, officials of the Royal Belgian Automobile Club set up timing equipment over a measured mile, and a party of English journalists came over with Lyons in a chartered DC-3 to watch.

Equipped with a set of metal pans to streamline the belly of the car, but with the standard windshield in place and convertible top erected, the XK 120 recorded a two-way average of 126.448 mph. With the top replaced by a small metal windscreen on one side, and with the other side of the cockpit covered with a tonneau, the speed rose to 132.596. The best one-way velocity was over 133.

This was sensational. As *The Autocar* expressed it: "By this bold demonstration of faith in the product, Jaguars [*sic*] have established their car as the world's

*"The view behind the wheel . . . brings the conviction that the designer was a shrewd psychologist," said* Road & Track. *"[T]he driver is mainly impressed with a feeling of unleashed power . . ."*

fastest unsupercharged catalogue model with full touring bodywork. Indeed, it is very doubtful whether any standard model in catalogue condition, even with the aid of a supercharger, has ever recorded such speeds over a measured distance."

The intrepid man at the wheel was a former race driver named Ron "Soapy" Sutton. His previous best-ever straightaway speed had been 112 mph in a racing car on the old Brooklands track. That had been something of an experience. But when driving his company's new street sports car 21 mph faster in Belgium, as he wrote later, "I was surprised to find I was less conscious of [the] speed."

Was the Jabbeke speedster indeed a "street" sports car? Enough so that it ran on 74-octane gasoline, normal for the era, drawn from a local pump. Enough so that it made a special demonstration pass by the assembled journalists at nearly walking pace—in top gear. And enough so that it went home to England the long way, engineer Wally Hassan and race driver Tommy Wisdom detouring down to the Alps to reconnoiter an upcoming rally.

Once back in Britain, the selfsame 120 was converted from left-hand to right-hand drive and entered in Jaguar's first postwar race, a one-hour event for production cars at Silverstone on August 20. It won, and a sister car was second. Incidentally, the winning driver that day was Leslie Johnson, the BMW 328 owner who had lent his car to Jaguar for investigation. He went on to become a very enthusiastic and successful factory-supported XK 120 racer—Britain's Old Boy Network in action.

That Silverstone victory was only the first in what would be a long skein of race, rally, and record successes for the XK 120. Among the most famous was the Dundrod Tourist Trophy of 1950, a 225-mile, three-hour event run in a fierce rainstorm on narrow, fast country roads in Northern Ireland. One of the new roadsters won after a fabulously forceful drive by young Stirling Moss, then just beginning his own great career.

In the rallying world, the most successful SS 100

driver had long been Ian Appleyard. When the factory released an XK 120 roadster to him, he promptly flogged it over 2000 miles of mountain roads to score the equivalent of an outright victory in the 1950 Alpine Rally. His navigator was his new wife, the former Pat Lyons. For both, it was the first of many wins in her father's sports cars.

As for speed records, there was the impressive week-long run in 1952 of a new XK 120 Coupe at the Montlhery track, just south of Paris. Driven by Moss and Johnson among others, in seven days and seven nights the car covered 16,851 miles for an average speed of 100.31 mph.

Not that all such outings were successful. At Le Mans in 1950, the hardest-driven of three Jaguars reached second place, but tore up its clutch. That same year, a four-car team contested the Mille Miglia, that arduous epic around the northern half of Italy, but the best they could do were fifth, Johnson driving, and eighth, Clemente Biondetti pressing on regardless of broken rear springs. Biondetti's stalwart attack on Sicily's wild Targa Florio ended with a broken connecting rod. The first 120s to race in America, home of tight road courses, tended to drop back from battles for the lead with severely faded brakes. Putting the best possible face on such failures, Jaguar's very tall racing manager, Frank "Lofty" England, pointed out that at least racing was "quickly showing up faults that might never occur at all in normal use."

Of course, his implication was correct. Speed competition never had been the prime purpose of the XK 120. Though William Lyons had initially referred to it as his "Super Sports" model, the entire thrust of his design was fast, enjoyable highway driving. And, after a painfully long wait, eager buyers (and writers) finally began getting a chance to try this new level of luxurious performance for themselves.

*Critics faulted the long clutch-pedal travel and the steering wheel's bullet-shaped hub. The pedals adjusted and the wheel telescoped, but there was no changing the counterclockwise-turning tach needle.*

CHAPTER VII

# XK 120: ON THE ROAD

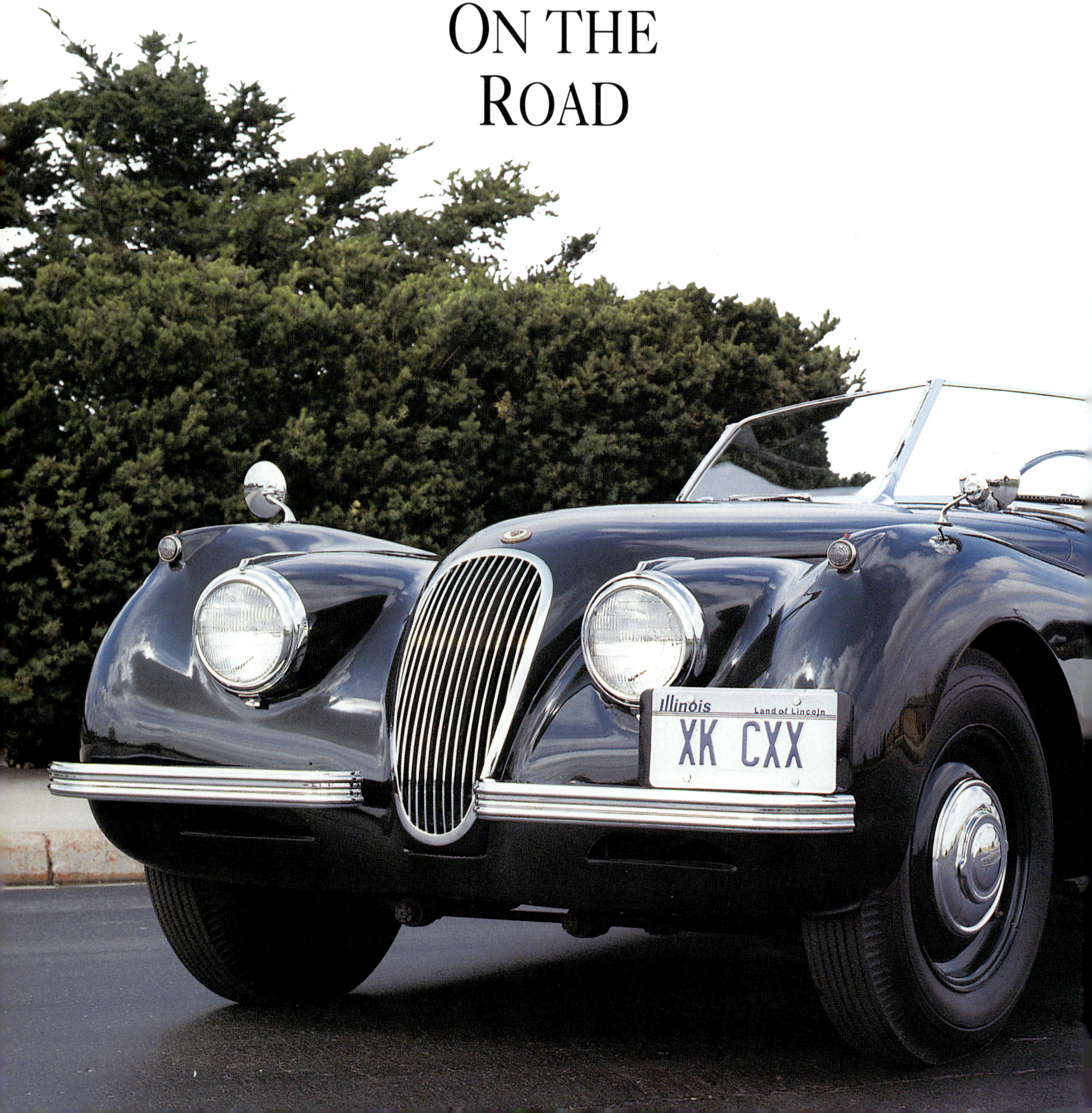

For 18 long, frustrating months after the October 1948 unveiling, examples of the wondrous XK 120 simply trickled out of the Browns Lane factory. For a lesser car, such a delay might have been crippling to its market prospects. But excitement was kept on the boil by the XK 120's inherent desirability, by the factory's carefully crafted competition program, and by occasional breathless magazine reports on the all-too-few aluminum-bodied roadsters released into private hands.

One of the first owners was that well-known Jaguar fan, Clark Gable. So eager was the actor to get his hands on the XK 120 he'd ordered that he happily accepted an invitation to "test drive" the first one in California—if only at the end of a rope: It was a show car without a crankshaft. When his own Jaguar-powered Jaguar finally arrived, he quickly put it through a personal evaluation routine, and reported his findings in a short article for the March 1950 *Road & Track*. His remarks are interesting as an indication of how revolutionary the XK 120 seemed to experienced enthusiasts at the midpoint of the automotive century.

"From Duesenbergs to and through hopped-up popular makes, I've owned and/or driven most of them. Many were fast but hard to handle on the turns; some lacked the acceleration . . . others were uncomfortable, uneasy, cumbersome, or otherwise undesirable from one standpoint or another. The XK has, so far as I've encountered, none of these drawbacks."

Gable lost no time in researching his new "Jag's" top speed on a California dry lake, and clocked it at 124 mph. "There isn't the slightest feeling of exceptionally high speed one generally has in smaller automobiles," he reported. "As for maneuverability, I'll stack the cornering abilities of the XK against anything I've ever driven. I have put her into as many types and kinds of slides as I know without once having the fear or uneasiness I generally have . . ."

*The XK 120 obviously was fast, but first impressions of hot-rod speed inevitably grew into an appreciation of the car's all-around competence.*

One of the earliest professional reviews of an XK 120 appeared in *The Motor* during November 1949. The British magazine's test car was actually the 1948 motor show prototype, later one of the trio of 1949 Silverstone racers, now updated to the latest production specs. Going straight to the same piece of Belgian superhighway scorched by "Soapy" Sutton six months before, the magazine staffers observed nearly the same two-way average speed: 124.6 mph. Both cars had been in the same trim, with normal windshield on, top up, and optional bellypan in place. Sutton's testing had shown the pan to be worth about 3 mph, at the cost of an uncomfortable rise in cockpit temperature. In November, nobody remarked on that. But *The Motor* did feel it worth remarking that the car was so stable at such rare velocities that the driver only needed one hand on the steering wheel.

As for acceleration, 0-60 came up in 10 seconds flat and the standing quarter-mile in 17. That, said *The Motor*, was "acceleration such as most drivers have never even imagined." Yet the report enthused about the engine's "docility," the clutch's "smoothness," and the Jaguar's "delightful willingness to crawl in tightly packed traffic."

On the negative side, the testers found they didn't much care for the seating, which lacked lateral support for hard cornering, or the steering-wheel position, which wasn't right for quick maneuvering. In fact, this did not seem a very wieldy sports car. "The car's precise controllability is not always fully appreciated at first," was the way they put it. However, just as with the old SS 100, they found they gained more confidence after a while, and eventually could hustle along winding roads at remarkable speeds with just fingertip control. But no amount of experience could compensate for headlights that were not nearly as bright as the performance.

XK 120s bodied all in steel, except for aluminum hood, doors and trunklid, began coming off the production line in April 1950. Though they looked very similar to the aluminum model, observant eyes did note some differences. According to marque historian Paul Skilleter, the best distinguishing point is that the all-aluminum roadsters had a rather bulky rubber grommet at the base of the windshield pillars, an area where the steel cars were much neater. The new bodywork also had subtle differences in the curves of some panels, and the headlight nacelles were set slightly lower. Inside, an earlier hump in the transmission tunnel for the starter motor was absent in steel cars.

*Factory illustration of the Drop Head Coupe, introduced in '53. Unlike the roadster, its conventional convertible top was water-tight and convenient.*

The interesting question of comparative weights didn't seem to much interest contemporary road testers. In fact, the switch to heavier metal doesn't seem to have made much of a difference to the car's performance.

Firm data is scarce in the literature, and sources vary in their opinions, but an aluminum roadster in street trim may have weighed somewhere between 2850 and 2950 pounds. (One report giving 2750 is frankly hard to credit.) Some experts say the steel version was a bit over 100 pounds heavier, but *The Autocar* thought it was actually lighter. That magazine tested a new steel-bodied roadster in April 1950, and listed its weight as 2919 pounds. This, said the report, was "some 40 lb." less than an aluminum car, one of the trio of 1949 Silverstone racers, that the staff had examined earlier.

But for some reason, this test of a steel production car yielded numbers far less impressive than those published earlier. This engine allegedly had the same 8.0:1 compression ratio, and hence 160 horsepower at 5200 rpm. But perhaps something else was wrong with the car or with the test method and/or conditions. Or perhaps this was a "real" car instead of a press-pool special. In any case, the published 0-60 time was a disappointing 12.0 seconds. *The Autocar* used words such as "stupendous" and "supreme" and "astonishing" about the Jaguar itself. But the acceleration it recorded was not even in the same league with the numbers recorded by its rival, *The Motor*, during its own test of an aluminum car five months before. Nor did it match the performance of the 3.5-liter prewar SS 100, which could reach 60 in well under 11 seconds.

On the top end, the 120 easily beat the SS, of course, thanks to its more modern aerodynamics. But without the benefit of wide-open Continental roads, *The Autocar*'s observed British Isles maximum was only 115 mph.

At the other end of the road-test credibility spectrum, in April of 1951, *Motor Sport* gave an obviously

steel-bodied roadster an empty weight of "approx. 25½ cwt." (2856 pounds) and a 0-60 time of 9.0 seconds. And this with an engine supposedly having the optional low-compression (7:1) pistons to suit Britain's dreadful 72-octane "pool petrol" and a consequent 150 bhp at a mere 5000 rpm.

But let us shed this fruitless fascination with firm figures and view the Jaguar through the eyes of its contemporary public. Everyone who drove the XK 120 in its youth was excited by its performance. Yet they found there was more to the car than its numbers.

"That immense turn of speed is almost by the way. It is not the essence of the car's charm, which really lies in the smooth flexibility of the engine, the light but precise steering, the comfort of the suspension, and the stability shown under all conditions of handling." So said *The Autocar* as it tried to put into words "certain marked characteristics which are superbly satisfying." Again we find reported an initial feeling of diffidence about the handling that gives way to enthusiasm with more experience. The body did roll in corners, but only so far, and one soon delighted in the "firm but light" steering and the way one could "hurl around bends with increasing confidence," tires screaming.

This, decided the magazine, was "a car for real motoring." It certainly offered a number of thoughtful driver-oriented details, including adjustable pedals and large, round, white-on-black primary instruments where "the positions of the needles can be seen in a flash." Similarly, "the secondary control knobs are not set in a confusing row, but are well apart so that the desired one can be found instantly." (Unfortunately, this sound ergonomic layout would be abandoned on later Jaguar panels.)

Wrote William Boddy in *Motor Sport*: "By the high quality of its finish and appointments alone the XK 120 represents very good value for money. Its very liberal speed and acceleration, accomplished with such willing ease, are unrivalled and to drive this Jaguar is to enjoy an

*The Fixed Head Coupe retained much of the roadster's rakishness and its closed cabin made the XK 120 a more accommodating long-distance tourer. Indeed, it was a Coupe that ran round a French track continuously for seven days and nights, covering a record 16,851 miles at an average of 100.31 mph.*

experience at once unique and embracing one of the highest pinnacles of modern motoring."

One began enjoying the experience immediately on settling into the leather seats. "The view behind the wheel of the Jaguar brings the conviction that the designer was a shrewd psychologist," bubbled *Road & Track*. "[T]he driver is mainly impressed with a feeling of unleashed power, engendered by the massive appearing hood and curving fenders."

Generally, people seemed pleased with the seats and driving position, although some did feel their legs were stretched out rather flat. Most seemed to take in stride the potential kink in the ankles caused by the vertical, almost backward, static position of the bottom-pivot pedals. But many did comment on the long clutch-pedal travel. There were occasional negative remarks about the big wheel, too. While the old SS 100's steering wheel had been adjustable for rake angle as well as distance, the XK 120's rake was fixed, and the rim was almost vertical in the hands. By turning a collar on the column, the wheel could telescope, but some taller (or wider) drivers never could find a happy compromise between the rim rubbing their thighs or the large "bullet" in its hub—a menacing style taken directly from Jaguar sedans—fouling their forearms on tight turns.

One or two test drivers said they found the backwards-reading tachometer confusing, although, of course, it accurately represented the engine's direction of rotation. To some drivers the hood appeared exceptionally long and maneuvering, accordingly, as cumbersome. In fact, the XK 120 had a tight turning circle of 31 feet.

The steering itself earned almost universal praise. Although the later adoption of rack-and-pinion steering on the XK 140 would make the 120's recirculating-ball system seem vague and unresponsive, in retrospect, when the latter was current, everyone thought it delightfully light, smooth and reasonably quick. Commentators often mentioned the ball-jointed front suspension, advanced for the day, in connection with the nice steering.

Similarly, before later advances showed how much better transmissions could be, almost everyone said they liked the 120's four-speed. The few gripes focused on

*The Coupe shared its engine and two-seat capacity with other XK 120s, but this 1950 study* (left) *shows Jaguar entertained notions of a four-seater.*

sometimes slightly stiff shift action and the non-synchronized first gear rather than the second-, third-, and top-gear synchromesh, which in later years would be regarded as poor. One actually didn't have to do much shifting with the broad-shouldered XK engine. The early cars had a very low, "stump-puller" first gear, and many drivers never used it. Starting out in second worked just fine, and someone in a lazy mood could then put the stubby lever directly into top—and leave it there for long journeys.

Brakes were one weak point that was recognized at the time. Because of the extra power, the higher weight, the enveloping bodywork, and the smaller drums required by its smaller wheels (16-inch versus 18), the XK 120 introduced to Jaguar motoring something no SS 100 driver had ever reported: brake fade. That was mainly a problem on the track, of course. On the road, it was easy enough to drive the car within this limitation.

Ride quality was obviously a matter of individual taste, as well as locality and habitual cruising speeds. Some people complained that springing was a bit on the firm side. Others enthused about how comfortable it was. And a few branded the car too soft because of the extremes of roll and pitch they observed. Even those with the last opinion, however, admitted that the soft springs seemed to allow the wheels to follow the road surface and maintain traction well. The only real flaws stemmed from the heavy live rear axle, which was located only by its long, limber leaf springs, and which at that early stage had no limited-slip differential available. Bumpy corners would make the axle hop sideways, and brutal acceleration could induce some tramp.

Handling also was a subjective matter, descriptions seeming to depend on the driver. Some said the Jaguar understeered, others that it was a neutral-handling car, while a few characterized it as an oversteerer. Everyone seemed to agree that the 120 was safe and controllable when driven quite hard, but anyone who described it as the cutting edge of the art may not have been going fast enough. John Bolster, a well-known racing driver turned journalist, nicely expressed the car's nature in *Autosport*: "The XK 120 will drift a bend in a delightful manner, but it resents being flung about . . ." He left us to guess if he meant he spun it.

Not that any of these little deficiencies cost any sales. The car's performance-with-practicality, its responsiveness-with-refinement, its elegance and beauty, and certainly its wonderfully alive, snarling exhaust note all had their own seductive power. Perhaps even some ardent, on-the-edge sports-car aficionados realized that they didn't always drive around everywhere at the screaming limit, and that Jaguar's style of motoring really was a pleasant way of covering the countryside.

As Bolster expressed it, "A long run in this car is a pleasure that is difficult to put into words. Whether it is its complete indifference to all kinds of road surface, its silence and smoothness, or the feeling of always having more power in reserve, I know not. Suffice it to say that the miles melt away without the slightest effort, and one never makes oneself conspicuous by sounding like a racer. For those who can afford the higher petrol consumption, the 3½-litre engine gives a contemptuous ease to high speed travel that no smaller unit can hope to match."

He liked the adequate, if not ample, luggage space, too. So did *Motor Sport*'s Boddy, who went on to defend this modern convenience to traditionalists: "[I]n case anyone asks what this has to do with real sports cars, let me say that the very rapidity and driving pleasure afforded by such cars make them appropriate for long distance touring."

All these were good and genuine qualities, and they built up a large and increasingly loyal XK 120 following, but it would be wrong to gloss over the real flaws, faults, and foibles that owners discovered with their Jaguars. While the 120 was a revolutionary automobile in many ways, it was by no means a perfect automobile, and once the first glow of easy speed faded, a more balanced view emerged.

In fact, troubles emerged. In no particular order, reported XK 120 problems included engine overheating at slow speeds in hot weather, weak clutch throw-out bearing and timing-chain tensioner, a water pump described

*The XK 120 blended high performance with durability and a degree of practicality unmatched by anything at its price. Its stoutness was demonstrated repeatedly by rallyist Ian Appleyard, shown above with his navigator and wife, Pat, daughter of Jaguar czar William Lyons. Unlike many of its contemporaries, the XK 120 had a sizable trunk nicely trimmed and even equipped with a light. Note the efficient spare-tire stowage design.*

as "notorious," occasional breakage of rear springs, cracking of disc wheels during racing, cracking of the aluminum hoods during normal driving and servicing, and occasional opening of these hoods at high speed—leaving at least one surprised owner with a perfect impression of the Jaguar's cat's-face hood emblem in his trunklid.

To its credit, Jaguar moved swiftly to correct many of these troubles, but others remained penalties of ownership. For example, some taller types, who had understandably bought on impulse like everyone else, sadly discovered that the cockpit was too cramped for them. Owners in warm climates learned that the footwells turned into ovens, making the hinged side vents introduced on later cars a popular retrofit. But that only worsened a dust-ingress problem experienced by those who had to use dirt roads. City dwellers found their low-slung aluminum doors too easily damaged by curbs. Owners forced to erect the roadster's folding top and side curtains saw that the assemblage not only harbored massive blind spots but was as ugly as a cheap tent, roared and rattled worse, and kept out the rain about as well as no top at all. (These complaints did not apply to the proper convertible, the Drop Head Coupe introduced in 1953, which earned universal approval).

Finally, early 120s had no heater, no defroster, not even turn signals—not even as an option. No air conditioning, either, of course; in the early fifties you couldn't get that even on open Detroiters. Besides, the sports-car buff would have thought it as ludicrous as automatic transmission, power steering—or a power top. As for audio entertainment, one was expected to ride in rapt silence so as to take in the full measure of the sensual symphony played by the twin-cam six.

If any of these difficulties turned anyone away, it didn't matter, because there was always a queue of clamoring customers for whom Jaguar had concocted, not to say invented, the ideal blend of automotive qualities. "Sports car" has always meant what an individual wanted it to mean anyway. William Lyons' definition suited a lot of people just fine.

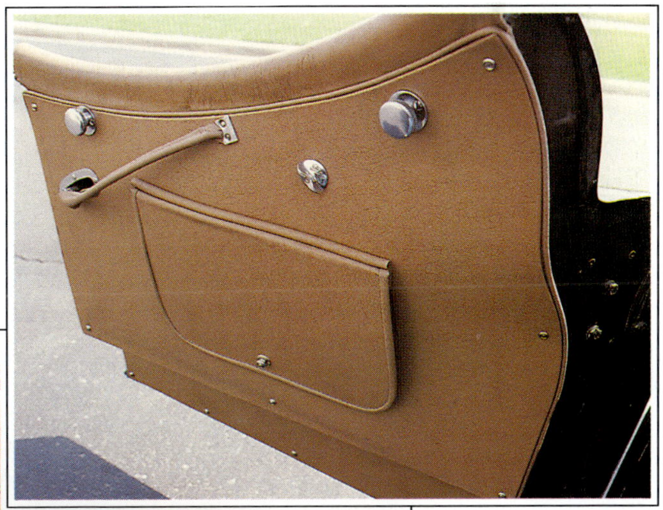

*XK 120's cockpit was classic sports car. Low-cut doors gave an exhilaratingly open feel, while the only audio entertainment was the engine's song. Taller drivers felt cramped, however, and the footwells could be toasters.*

*Poor rearward visibility was one penalty of the roadster's fussy top, while refinement and flexibility were hallmarks of the 3.4-liter six. First gear was a stump-puller, so some drivers started in second and went right to fourth.*

Yes, the XK 120 was first and foremost a sports-touring car. But for those of its customers determined to treat it like a sports-racer, Jaguar drew from its own ongoing competition experience and offered a set of recommended high-performance parts and official tuning advice as early as mid-summer 1951.

Given the availability of fuel of suitable octane (85-90), one could fit 9.0:1 pistons, a higher-lift camshaft, different needles in the SU carburetors, colder spark plugs, and a low-hanging dual exhaust system with straight-through mufflers. That would show about 190 bhp at 5400 rpm. To take proper advantage of it, one also was supposed to install a special crankshaft damper, a lighter flywheel, and a new, balanced clutch with stronger center plate. Removing the thermostat would help with engine cooling. Four different rear-axle ratios were offered by the factory.

For the chassis, Jaguar could supply 20-percent stiffer front torsion bars and rear leaf springs. For the body, there were lighter bucket seats available, and "racing windscreens" with cowlings to replace the big V-shaped windshield. Even the tiny rectangular central mirror—the only mirror on 120s—could be faired-in.

For the brakes, alas, there was nothing but slightly thicker shoe linings and the pathetic suggestion to remove the hubcaps and rear wheel spats on disc-wheeled cars to promote airflow over the drums. Some owners did attach cooling scoops to their XK 120 brakes, but discovered that they were as good at scooping in rain as air. Others had better results from installing wire wheels, which were thought to allow more air circulation around the drums as well as easier escape of heat in radiant form.

These hop-up parts proved popular, partly because they proved so practical for everyday use. This prompted Jaguar to offer many of them—plus a set of Rudge-Whitworth knock-off wire wheels—in a "Special Equipment" package for later models, a nice example of how racing really can improve the breed. In the United States, where "modifieds" were all the rage in sports-car racing, these pumped-up 120 SEs were given the suffix M instead.

A U.S.-spec XK 120M tested in 1953 by *Road & Track* had 8.0:1 pistons and was rated at 180 bhp at 5300 rpm. The torque value was 203 lbs/ft at 4000. The car hap-

*Engine overheating in hot weather and, in extreme use, brake fade, were chronic XK 120 shortfalls. Critics debated its handling: some described understeer, others oversteer, still others found it rather neutral. "Perhaps the real charm of the Jaguar," concluded* Road & Track, *"lies in its response to the driver's skill and proficiency . . ."*

pened to be a Fixed Head Coupe, and its empty weight as given was a suspiciously even 3100 pounds, distributed 1500 front, 1600 rear (48/52 percent). With an additional 335 pounds of personnel and gear aboard, and an experienced professional at the clutch pedal, the car was clocked at 8.5 seconds for the 60-mph sprint and at 16.66 in the quarter. From the accompanying graph, trap speed was about 82. Top speed was 120.8—at 3.77:1, the axle ratio was a bit shorter than the standard 3.64. Overall miles per gallon worked out to 15.2.

At $4460, the M cost $395 more than the regular XK 120 coupe. Porsche's 356 Super also ran about $4400. The Chevrolet Corvette introduced that year was pegged at $3513.

"Simply tremendous," said *R&T* of the M's performance. "The Modified Jaguar will outperform most production cars in the world today." Yet nothing had been lost of the XK engine's renowned flexibility, for even in its M-iteration it would pull smoothly in top gear from 15 mph to 120.

"Perhaps the real charm of the Jaguar," *R&T* continued, "lies in its response to the driver's skill and proficiency, for this is a car with a dual personality." One could "putter about town in high gear" or enjoy the "brilliant acceleration" through the gears. Again, despite its stiffer suspension, the 120M was a comfortable and stable highway cruiser, even in windy conditions, but it had ideal sports-car handling in the mountains. "The slight weight bias at the rear of the Jaguar coupe shows up well on winding roads, there being neither over or understeer, up to the point where the rear end finally 'breaks away' . . . This car has no vices, even for the novice sports car driver."

In an American publication called *Auto Sport*, writer Barney Clark extolled another sports-car virtue of the 120M: its "Wagnerian music." The "truly noble, soul-satisfying trumpeting that ensues when you depress what the British so aptly call the 'loud pedal,'" he said, was reason enough to buy the car. "It is an utterly distinctive sound—hard, taut, competent; the voice of a really potent high-performance sports car."

Clark did have the goodness, however, to warn his readers that the decibel level "rings 'General Quarters' for every cop in the Western Hemisphere."

CHAPTER VIII

# C-TYPE: RACING TO IMMORTALITY

*Encouraged by the showing of only moderately modified XK 120 roadsters at Le Mans in '50, Jaguar launched an all-out assault in '51 with the XK 120C, C for "competition." It retained a number of stock XK 120 parts, but the production twin-cam 3.4 six was tuned for 204 bhp, and the car got a new space-frame chassis, redesigned suspension, and fresh aero bodywork. Known as the C-type, it won its very first race, igniting Jaguar's run of five legend-making conquests of the French classic in the fifties.*

Why do so many so love the Jaguar? Because of something that, according to legend, William Lyons said one summer's evening in France in 1950. He said "Yes."

Actually, we can suppose he said something more like, "Right, then, let's have a ruddy good go at it!" His British sporting blood was up, you see. He'd just spent 24 hours watching a trio of his still-new XK 120 "Super Sports" roadsters competing in the Le Mans "Grand Prix d'Endurance," the greatest sports-car race on earth. They were virtually off-the-shelf street cars with only minimal racing modifications, and the entire program was no more than an investigation. Yet two cars had finished. That was honorable and encouraging. One of them had shown the potential to win. That was . . . exciting, exciting enough to authorize something as unnecessary as it was grand: a full-scale attempt on outright victory the following year.

Under usual circumstances, Lyons was cautious, even wary, of formal participation by his factory in motorsports competition. He saw more to lose than to gain. Jaguar Cars, Ltd. was doing well, building up a following, making a profit. More orders had come in already for the sleek, spectacular, sensual XK 120 than could be filled for a very long time. Yes, mass production had just started on the steel-bodied version, and the impatience of the customer list was finally beginning to be sated, but the car certainly didn't need any boosting.

Nor, really, did the company. Jaguar's reputation was now established. A fine new modern sedan was about to be launched, the Mark VII, with the same superb XK engine so widely desired in the sports car. It could scarcely fail. No, there might have been many sound arguments made against spending money and man-hours to risk a very public loss on a venture as precarious as an endurance motor race. Happily, William Lyons had no need of harking to any negative counsel. Jaguar was his company. He could jolly well have it do as he pleased.

Going racing at Le Mans must have pleased most of the people on his staff. As was Lyons himself, many were motorsports enthusiasts and had some personal experience at it. His chief engineer, William Heynes, was keen on racing and dabbled in it as occasion permitted. The firm's service manager-cum-racing team manager, Frank "Lofty" England, had a deep background in the sport. There were others. In a careful, scientific, low-key sort of way, Jaguar Cars, Ltd. was something of a hotbed of auto racing fervor.

Yet the ambition was kept on the firm reins of judgment. Lyons agreed to go racing only when he saw his company could field a successful effort. Every step, though driven by emotion, was securely planted on logical methodology.

When Lyons invited the world to watch an XK 120 exceed 132 mph at Jabbeke in May 1949, his henchmen first took great care in private to be sure the thing could be done. Similarly, before the identical car and two of its fellows were entered for the 120's first circuit race that August, Lyons himself participated in preliminary testing. When the Jabbeke car won at Silverstone, it was no real surprise.

Doing so well at Le Mans the following June was a little surprising, but only a little. By then, the XK 120 had begun to prove itself not only fast but durable. And the three cars entered in the 24 Hours were prepared with care and common sense. No attempt was made to squeeze more power out of the engines, or to abandon what had been working in other respects.

Bumpers were stripped off, and the cars did have the low, narrow aeroscreens and form-fitting bucket seats offered as racing accessories. But all ran without the belly pans used at Jabbeke, and their aerodynamics were further compromised by pairs of auxiliary driving lights. To preclude the real chance of the long, rear-hinged hoods blowing open at speed, the mechanics added leather straps. Quick-action caps were installed on large, 24-gallon-Imperial gas tanks (28.8 gallons U.S.) that replaced the standard 14-gallon ones (16.8 U.S.). Since these took up most of the trunk volume, spare parts and tools were stowed in the only remaining vacant space, the passenger seat.

The greatest departure from stock specification had to do with brake cooling, which was going to be important for minimizing fade at the ends of the long, long Le Mans straights. Rather than buy aftermarket Al-Fin drums, which Jaguar feared might crack in such hard service, the team chose to try to increase cooling airflow by sandwiching discs with small vanes between the normal iron brake drums and the steel disc wheels. For good measure, they also drilled strategic air holes into the drums themselves.

All three cars started well, and mixed it up with the leading pack for some time. At the end of the first hour, the white 120 driven by Leslie Johnson was holding fifth place. He and his co-driver, Bert Hadley, continued pushing as hard as they could all night, and actually worked up to second place overall by noon the next day. At that point they were lapping at a speed that by the finish could have reeled in the leading car, a 4.5-liter Grand Prix-based Talbot. However, in their concern to avoid overheating their brakes, which still were prone to fade despite the modifications, they were downshifting the transmission too brutally. With just under three hours left in the race, their clutch center tore apart. Johnson's roadster coasted to a helpless stop within sight of the pits. The other two cars were driven more conservatively, and though neither was trouble-free, they did keep running and came home twelfth and fifteenth.

It's unlikely the Talbot team would have let the Jaguar beat it to the checkered flag. But to the men from Coventry, it was obvious their cat had all the inherent heart and muscle needed to win this race. If what amounted to a stone-stock street sportster could lap at almost 97 mph, just 5 mph short of the lap record established by a Talbot GP car wearing lights and fenders, just imagine what the mighty XK engine could do in a proper racing chassis. Heynes may well have shown some quick calculations and sketches to the Old Man during the race. Anyway, in the heady flush of euphoria soon afterwards, Lyons said "Yes."

Work on what would become the 1951 C-type did not begin immediately. First, the Mark VII sedan had to be finished, introduced, and sent properly down the production line. There were also a number of people racing and rallying the XK 120 whose needs had to be attended to. But in the autumn, Heynes and his backroom boys finally turned to the joyous job of designing their Le Mans car.

Cars, actually. Realizing the magnitude of what he'd agreed to, Lyons set in motion a less-ambitious backup program. In case his trio of radical racers wasn't

*The C-type was among the very few Jaguars not styled by William Lyons. Aircraft aerodynamicist Malcolm Sayer designed its wind-cheating bodywork, here shown in an early clay-model rendering.*

*A more detailed wood-carved scale study emphasizes the C-type's competition bearing with its smoothed-over passenger side and its small racing windscreen.*

done on time, there would be a trio of cars that looked a lot more like XK 120s, albeit with lightened street-car frames and handmade magnesium bodies. These so-called LT bodies were built but never had to be used at Le Mans. They wound up on standard chassis in the hands of privateers.

With the primary race car the basic idea was to "repackage" as many standard XK 120 components as feasible into a smaller, lighter, more aerodynamic bodyshell. But the racer would be no simple hot rod. It would represent a thorough, rigorous application of cold science.

The team began as would any of us, with thumbnail sketches. One of these, perhaps by Heynes himself, has survived to be published by Philip Porter in his *Jaguar: History of a Classic Marque*. It clearly shows the essential elements of the final car: wheelbase reduced by six inches to 96, a 40-gallon (48 U.S.) fuel tank mounted over the rear axle and extending into the tail, and a spare wheel laid horizontally under the tank extension. The driver perches just ahead of the rear wheels, his body flexed a bit to nestle between there and the engine, which is carried well aft of the front-wheel centerline.

After a series of trial constructions in paper and wood—even broomsticks, said Heynes years later—and a lot of tedious math in that pre-computer era, the racer took form in a beautifully logical way. As had been hoped, many stock parts were deemed suitable, but as many were replaced with special fabrications. The chassis was now a nest of steel tubes ranging between one and two inches in diameter, all welded together to make up a very strong but light "space-frame." Additional strength came from incorporating some sheet-steel panels in the firewall and rear bulkhead, making those structures virtual monocoques.

The chassis essentially ended just behind the cockpit, with the rear axle hung off the back on trailing links. These links were attached at their forward ends to the springing medium, which was a single transverse torsion bar, fixed in the center so it acted like two separate bars. The advantages of this design were lighter weight, more positive location of the axle, and elimination of the inter-leaf friction that constituted an unpredictable damping force with the roadgoing 120's conventional semi-elliptics. Instead of the lever-type shocks at the back of the street car, the competition model used telescopics.

One other interesting innovation appeared at the rear. To help prevent wheelspin in this era when limited-slip differentials were rare, Jaguar designed a torque link that mounted atop the right side of the axle, leading forward. Normally, in a conventional beam axle, the pinion gear has a tendency to "climb up" the ring gear. This creates a lifting moment that reduces the weight on the right-rear wheel, allowing it to spin just when engine output is highest. That's why powerful cars with this type of rear suspension and no limited-slip often leave just one black streak of rubber when scorching away. Jaguar's novel link was placed and angled in such a way as to counteract this lifting force and press the right-side wheel back down onto the road. Of course, it also located the axle in a rotational sense, thus reducing any tendency to tramp or judder on acceleration. On the original design, this link was A-shaped and had the additional duty of locating the axle laterally against cornering forces. At 51 inches, rear track was wider by an inch compared to the street car.

At the front, the control arms, longitudinal torsion bars, and transverse anti-roll bar were at least similar to those on the XK 120, though the geometry was a little different, while at 51 inches, the track was the same. The steering was completely different, a rack-and-pinion system replacing the street car's box-type recirculating-ball

mechanism. This was adopted mainly in the interest of better road feel, but it also brought a quicker ratio, and thus faster helm response.

The brakes were still iron drums all around, but at the front was a self-adjusting mechanism to take up wear as it occurred. This prevented the disconcerting long pedal travel that would build up in early XK 120s in sustained hard braking. Wheels were knock-off wire types of the same 16-inch diameter as the road model's, but the rims were an inch wider (six inches in all), made of aluminum, and mounted Dunlop racing tires.

The bodywork, entirely new, was a simple low, rounded envelope designed to cheat the wind as well as that black art was then understood. In fact, the shape was owed to a professional aircraft aerodynamicist, Malcolm Sayer, making the C-type one of the few Jaguars that would not be styled by Bill Lyons. The reason, as Bill Heynes joked years later, was that the boss was away on a sales trip to North America during the critical weeks—which also allowed the work to be done in record time.

This body was purposeful in other ways. A single right-side door bespoke the no-frills, pure-speed intent, and the entire nose section pivoted up on front-mounted hinges for easy engine access pitside. A passenger seat was provided, but it was only a hard, token affair high atop a toolbox, and getting into it demanded a careful climb over the door-less—and fragile—aluminum bodyside. There was no trunk (no need for it), but a trap door was set low down on the tail for spare-wheel access.

In contrast to the year before, the XK six was treated to a mild hop-up for its 1951 racing chores. Larger exhaust ports served larger exhaust valves, and the camshafts lifted all the valves by an extra $^1/_{16}$ inch, $^3/_8$ inch total. New valve springs lifted the designed valve-bounce revs to 6500, a healthy 1000 rpm above the peak-power speed. New pistons with solid skirts increased compression ratio to 9.0:1, but the pair of SU carburetors were the regular production items with 1$^3/_4$-inch bores. Cold outside air was ducted to the carbs from the nose, rather like the system on the street 120 but without the air cleaner. The exhaust manifold remained a casting, but it had different dimensions to suit the higher working revolutions. There were better crankshaft bearings, and a stronger crankshaft damper. Of course, the flywheel was lightened. The distributor was revamped, too. At 5500 rpm, the race engine put out about 204 horsepower.

Well learning from the Johnson/Hadley failure of 1950, Jaguar engineers installed a stronger clutch with a solid center. While at it, they switched to a different transmission shaft to make it easier to change ratios. The same consideration led them to exchange the road car's ENV rear axle for one made by Salisbury, which offered a greater variety of final-drive gearing.

What to call it? At one stage some thought apparently was given to "XK 150," after its projected top speed. In the end, despite its many departures from the road-going 120 (some of which would later be adopted for street use—more "racing improves the breed" from Jaguar), it was still referred to as an XK 120, although a "competition" version: XK 120C. That and chassis numbers beginning with the letters XKC made C-type inevitable, and it stuck.

Race-ready, the XK 120C was much lighter than its street sister, coming to about 2100 pounds. That was according to figures published by Jaguar engineer Robert Berry in the book *Jaguar: Motor Racing and the Manufacturer*. It represented a difference on the order of 30 percent. Whereas the road car carried more than half its

*The nose section pivoted forward to reveal the engine bay, and a small right-side door was the only cockpit access* (opposite). Left: *C-type No. 20 rips through a turn on its way to victory at Le Mans in '51. Broken oil pipes sidelined two sister cars, but Peter Walker and Peter Whitehead brought this one home at record speed.*

weight on the rear wheels, the racer was slightly nose-heavy at 51.5-percent front/48.5-percent rear. Its body was only slightly smaller in frontal area compared to the XK 120 roadster equipped with aeroscreens, but sufficiently streamlined as to require 22 percent less horsepower to maintain a given speed. In several tests, the original C-type topped out at nearly 144 mph.

The three Le Mans cars were completed only about six weeks before the 1951 event, but that was time enough for Stirling Moss and Jack Fairman to do some testing. They uncovered little that needed changing. Jaguar's first racing job had been done well.

So well that the C-type won its first-ever race, the race for which it had been designed, the most important sports car race in the world, the *Vinqt-quatre Heures du Mans*. Beating Allards, Aston Martins, Cunninghams, Ferraris, Nash-Healeys, and Talbots, a Jaguar driven by Moss came from a mid-field start to take the lead on the third lap, and eventually broke the lap record by six seconds, at 105.2 mph. (He later said he could have lapped at 107.) The other two C-types were going well, too, and in the fifth hour the novice team was running first, second, and third.

But then, two engines were cut off at the knees by broken oil pipes. Greatly worried, management ordered the remaining car to slow down. Luckily, the strongest of the opposition had also faded, and the Jaguar driven by Peter Walker and Peter Whitehead was able to stroke home to victory at 93.49 mph, beating a Talbot by 67 miles. It was the first British victory at Le Mans since Lagonda had managed the feat in 1935.

C-types went on to score two more victories that thrilling 1951 season. Moss copped one in Ireland, where he'd won the 1950 Dundrod Tourist Trophy in an XK 120. He triumphed again at the southern England circuit of Goodwood. Three events, three wins for the new Jaguar.

Great plans were laid for 1952. For one thing, a series of production C-types was made for sale to private owners. And behind the scenes, Jaguar entered a long and demanding program with Dunlop and Girling to develop a racing disc-brake system.

The disc brake was not a new idea even in those distant days. Englishman Frederick William Lanchester had patented an embryonic design for his car way back in 1902, and many others experimented with various ideas for both racing and road use into the early forties. Girling and Dunlop were among the pioneers, as were Lockheed, Goodyear, General Motors, and Ausco-Lambert in America.

Postwar progress, made largely through liberal borrowing from aircraft technology, prompted Chrysler to offer the Ausco-Lambert "Safety Brake" as an extra-cost alternative to drums on its large, low-volume Crown Imperial sedans starting in 1949. The following year, tiny Crosley of Cincinnati, Ohio, made Goodyear-Hawley "spot discs" available as a regular factory option for its snazzy little Hot Shot roadster. Alas, these weren't thoroughly tested and suffered frequent sticking from salt corrosion. A few disc-equipped Hot Shots may have been raced, but it's likely that the first serious attempt at using discs in competition was Harry Miller's radical four-wheel-drive Indianapolis racer way back in 1940.

European motoring writer Jan Norbye, in a 1973 retrospective for *Special-Interest Autos*, records that the modern disc brake was born in postwar England at the Dunlop Rubber Company: "While Goodyear based its aircraft brake on Hawley patents . . . [Dunlop] created its own . . . which in turn produced a number of patents for

*Seeking more top-end speed, Jaguar tried "droop-snoot" bodywork for Le Mans in '52, but aerodynamic lift and poor engine cooling brought back the original C-type design* (left) *for '53. Note the change from large fender vents to a series of slats.*

Dunlop. The central feature of the Dunlop disc brake was that both rotor and caliper were fixed in the axial plane. Though needlessly expensive by American production standards, it gave good results. Girling bought a license to produce passenger-car disc brakes under Dunlop patents, and the prototype unit was exhibited at the London Motor Show in Earl's Court in 1951."

Jaguar was among the first automakers to test the Girling disc, which looked very promising, at least according to Norbye's first-hand recollections: "I remember driving a Jaguar Mark VII with experimental Girling disc brakes at Goodwood in 1952. It was very impressive to haul the big sedan down from 110 mph to standstill in about 300 feet with perfect lateral balance." That was apparently in the dry. Had it been raining, Norbye would surely have commented on the system's wet-road performance: superior to that of even the best drum brakes. It was superior fade-resistance, more than wet-weather efficiency, that made discs advantageous for racing, however. Though Jaguar naturally wanted this competitive edge for Le Mans '52, the testing wasn't completed in time, so the brakes didn't make it.

But an improved engine did. Compression ratio was reduced to 8.5:1 to better cope with the poor-quality gasoline supplied by the French race organizers, and the torque curve was fattened up with larger two-inch carburetors. The combined effect was better acceleration despite the loss of a single horsepower at the top end. Then, at the last minute, the company decided to run completely new bodywork. It would prove a disaster.

The purpose was sound enough: more top speed. So was the motive: new competition. The instigator was Stirling Moss, just returned from the Mille Miglia in Italy (where he had been race-testing disc brakes, incidentally), reporting that his C-type been blown off by the new Mercedes-Benz 300SL coupe, a car that also would run at Le Mans. Moss pressed his case so emphatically that better judgement was abandoned. Thus were radical new aerodynamic shapes drawn up and built, and the trio of cars went off to France without any testing—and without any backup in the form of proven cars from the previous season.

There was time enough for practice laps, though, and the new "droop-snoot" Jaguars proved faster down the Mulsanne Straight by some 10 mph, about 153 all-out. But there were two problems.

One was the new shape. Low and rounded at the nose, low and pointed at the tail, it not only looked like an airplane wing (in cross-section) but acted like one. The cars wanted to fly. In fact, their rear ends lifted so much that load on the rear tires was reduced by a good quarter, and the drivers came back to the pits ashen-faced to report evil instability. The older body had shown a tendency to lift a bit, but nothing like this.

The other problem was even worse: overheating. Within a few laps of practice, all three cars started boiling like teakettles. And the damage done was apparently permanent, compounded by an evident lack of any spare engines. Although two radiators were hastily modified before race time, all three cars retired in ignominy just an hour after the start. To William Lyons, it must have seemed the realization of his blackest dreams.

Ironically, the new Mercedes coupe showed nothing like the straightaway speed that Moss had feared. But it won.

As later tests proved, the new C-type's overheating problem was not caused directly by its aerodynamic shape but by insufficient water flow in a new cooling system hastily designed to fit under the drooped snoot. But this, at least, was easily solved. At that stage of aerodynamic know-how—and the company's limited resources—the high-speed instability seemed inherent and incurable. It would be another decade before American driver Richie Ginther, during tests with Ferrari, would invent the trim tab, or "ducktail spoiler," to tame such handling quirks.

Jaguar didn't give up its low-drag dreams after Le Mans '52, but it did scrap the droop-snoot and revert to the proven original C-type body while making a firm resolve never again to go racing in haste. Later in the year, Moss provided a history-making bright spot by making haste at Rheims to give disc brakes their first-ever competition victory.

That '52 season saw other successes—and other failures—all with special lessons to teach. The Jaguar team thus contemplated a fourth visit to Le Mans with a lot of experience under its collective belt.

The 1953 C-types reflected this with a further increase in power. There was another change of carburetion, to a trio of two-barrel, 40-mm Webers, which were fed not from the nose but from a scoop atop the hood. These and other alterations resulted in a little more top-end power, up to about 220, but at lower, safer revs, 5200. The extra power gave back half the speed lost to readopting the original body, and the best of the cars was clocked at nearly 149 mph. Even more useful was a further swelling of mid-range power for yet better acceleration. Additional detail improvements included a stronger crankshaft damper; new, better-sealing piston rings; a stronger but smaller-diameter triple-plate racing clutch; and a water pump and radiator core redesigned for better water flow.

Though the '53 cars looked a lot like the original '51 C-types, they were more than 50 pounds lighter. Some of the weight-saving came from lighter-gauge frame tubing, some from thinner body metal, some from painstaking detail changes to things like the electrical system. Even the main battery cables were now aluminum, rather than copper, a common aircraft trick. Also borrowed from aviation were rubber-bladder fuel cells, which were not only safer but lighter and more durable; the previous aluminum tanks had sometimes split in tough races.

Racing breakages also led to a change in the rear suspension, where the axle torque-link, originally A-shaped, was now just a simple bar because it had been relieved of its task of lateral location. That job was now handled by a more conventional Panhard rod, running from the right side of the axle to a bracket on the left side of the frame.

And then there were the disc brakes. The R&D program had been long, difficult—and sometimes exciting when test drivers would occasionally arrive at a corner with no braking ability at all. But the early problems of pedal effort, pad knock-back (improper seating in caliper), and fluid boiling had finally been solved and the relentless, fade-free power of the servo-boosted disc brake played a major role in Jaguar's second Le Mans victory.

This was harder-fought than the contest of two years earlier, the Jaguars facing factory efforts from Alfa Romeo, Allard, Aston Martin, Cunningham, Ferrari, Frazer-Nash, Gordini, Porsche, and Talbot. Little had been expected of the C-types, not only because of their early retirement the year before, but also because of a string of more recent failures.

The weekend even started on a distinctly unpromising note, officials objecting to Jaguar numbering its fourth, practice car the same as one of its actual racers. That was considered a serious infraction, and the drivers involved figured their car was disqualified. Tony Rolt and Duncan Hamilton, two of the larger-than-life characters typical of motorsports (particularly in Britain), went off to enjoy Friday night in their own way amidst the giant party that was, and is, Le Mans. But William Lyons managed to mollify the officials with a fine of 25,000 francs, and got the car reinstated. The only remaining problem was the condition of his two wandering drivers, who'd not bothered to get any sleep.

Come Saturday afternoon, Moss passed a brutal, big 4.5-liter Ferrari for the lead on the fifth lap, only to pit with fuel-system troubles. But Rolt and Hamilton, driving on adrenaline, took up the slack. There ensued an hours-long battle at tremendous speeds, during which Hamilton ran into a bird that smashed half his aeroscreen on the way to making his nose hurt as much as his head. But the cars and crew from Coventry basically ran the expensive, exotic competition into the ground. At four o'clock on Sunday afternoon, Rolt/Hamilton came across the finish line first, with Moss/Walker second and Whitehead/Ian Stewart fourth. The winner's average speed was a new record, 105.85 mph.

This was epic. A second triumph on the very ground made holy to every Briton by five Bentley victories (1924-30) positively cemented Jaguar's stature in the sports-car world. To make sure the world-at-large knew

*Cutting weight and upping horsepower to 220 made the '53 C-types quicker; top speed was nearly 149 mph. But the perfection of disc brakes proved the weapon Jaguar needed to beat the likes of Ferrari, Porsche, and Alfa Romeo for its second Le Mans victory, an epic win at another record pace.*

about it, the company's ever-alive publicity department had the bright idea of sending a telegram to the newly crowned Queen Elizabeth II and leaking it to the press. It read:

THE JAGUAR TEAM HUMBLY PRESENT THEIR LOYAL DUTY TO HER MAJESTY AND WISH TO ADVISE HER THAT IN HER CORONATION YEAR THEY HAVE WON FOR BRITAIN THE WORLD'S GREATEST INTERNATIONAL CAR RACE AT LE MANS, FRANCE, YESTERDAY.

The recipient was good enough to have her private secretary reply in kind:

THE QUEEN WAS VERY PLEASED TO LEARN OF THE SUCCESS OF THE JAGUAR TEAM. PLEASE CONVEY HER MAJESTY'S SINCERE THANKS TO ALL MEMBERS OF IT FOR THEIR KIND AND LOYAL MESSAGE.

But the C-type was now finished as a factory racer, though it would continue for several years as a popular and effective weapon in the hands of owners all over the world. It was even used by several people as a real "Super Sports" street car. Among them was none other than Dr. Giuseppe Farina, the Alfa Romeo Grand Prix ace.

Two others were staffers at *The Motor*, which in late 1952 sent them off in a C-type for the obligatory (and immensely enjoyable) Continental road test. This was a 200-bhp production racer with two SU carbs, 8:1 pistons and a 3.31:1 final drive ratio. "Unladen kerb weight" was given rather casually as "20 cwt." or 2240 pounds, and the front/rear distribution was supposed to be precisely 50/50.

To the test team's surprise, they found enough room to stow their belongings in the hollow body either side of their seats. To their relief, they found the rush of air over the screens kept rain out of the cockpit (a valid excuse to keep the right foot down). To their satisfaction, they found the car was entirely happy in slow-speed traffic and on wet Belgian paving blocks and tram rails. Yet, this genuine Le Mans racer would cruise comfortably at 120 mph, would rip on up to 135 any time the opportunity presented itself, and when given a long-enough clear road would top out at a timed average of 143.7 in rainy conditions described as "very adverse." At such uncommon velocities the car tracked absolutely true, although above 130 there was "a curious sense of becoming faintly airborne."

During a moment of dry weather, this C went 0-60 in eight seconds flat and covered the quarter in 16.2 at 90 mph. Overall gas mileage worked out to 13 per U.S. gallon. The price in English pounds was £1495, pre-tax. (U.S. list price was $5860.) That was a bit stiff for journalists in those days, but all in all they allowed as how they'd be happy to use the racing Jaguar as everyday transportation. It was a "thoroughbred," they said, "a docile and tractable machine completely without temperament." Their 500 miles of hard driving in Europe was "a great and memorable experience." Sigh.

Climbing aboard a Jaguar XK 120C takes some dexterity, because the single door on the right is very small, and the nest of space-frame tubes makes quite a high sill. Once seated, space is at a premium, especially fore/aft. The 17-inch, three-spoke plastic steering wheel feels large, upright and close, rather like a yacht's. As in

*Retired as a factory racer after the '53 Le Mans win, the C-type was a track success in the hands of private owners for several more years. It also made a stunning "Super Sports" road car. No windshield wipers, no top, no heater, just raw performance: 0-60 mph in eight seconds and a 140-mph-plus top speed. The selling price in 1952 was $5860. Jaguar built just 54 C-types, but that's all it took to achieve sports-car immortality.*

the street car it can be adjusted fore and aft, but in this case a wrench is required. There is no room to recline the bolt-upright seatback, even were it capable of doing so. The footbox does not extend forward of the firewall, so your knees seem a bit too bent. Although the body is almost exactly as deep as the roadgoing XK 120's, there is no frame to sit on, so you feel comparatively submerged.

Looking around, you see a lot more metal than upholstery. For wind and weather protection you rely on a pair of fold-down aeroscreens or, on some cars, a single low sheet of thin plastic. But that's it: No wipers, no top. A heater? It's the big iron stove up front. With the rounded tail full of fuel tank and spare wire wheel, the only possible place to stow any luggage is under your elbows.

This Jaguar, you are forced to realize, is not a car for long-distance comfort. Just long-distance pleasure. The C-type gives a very satisfying sense of being nestled amongst serious, no-foolin' machinery, stark as an old motorcycle. It is by no means intimidating or unpleasant, just alien to the coddle-the-customer philosophies of conventional automobiles.

Though tuned for racing, the big six awakes easily at a push on the panel button and has surprisingly good low-speed manners. If you don't ask for too much with the throttle at low rpm, which can set up a stutter, it pulls with almost the same docile strength for which the normal XK is famous. But where the street engine keeps on pulling that way right around the tachometer—a rush of smooth torque without hump or hollow—the C-type comes to sudden, vivid life at 3000. It then eagerly takes full boot. Raucous and racy, the sound from the exhaust becomes pure music, and the engine wants to rush to the redline with joyous abandon.

Perfectly complementing the symphonic nature of the engine are those sculptural curves stretching way out ahead. Driving a C-type fast is almost like attending a ballet, the eyes feasting along with the ears. While you sit low enough to worry a little about where the corners are in traffic, on the open road this is no problem.

The clutch is light and sweet, and its characteristic long travel is no real bother, but the gearshift at your left has a surprisingly long and lanky travel, and getting it all the way over into the reverse plane (left and forward) takes a practiced backhand whap. And at least when the box has some miles under its synchros, it protests overly quick shifts with irritable grating noises.

By contrast, the rack-and-pinion steering is a pure delight, having that smooth, light, precise feel of perfectly honed vintage machinery. In an opposite way, the car itself feels a bit vintage on the road, somewhat stiffly sprung and a trifle darty over bumps. Although it seems a large car to the eyes, it feels a small, nimble car in the hands, and can be hurled around with ease. In very tight corners, it takes a surprisingly delicate foot to keep the rear wheels from juddering on braking-in, and from spinning on powering-out. On more open turns, the C-type handles and sticks well for its period, but there is something in its chassis that likes to make a series of choppy arcs out of long sweepers. Word was, drivers who did well with Jaguar's first pure racer had learned to bend it to their will with decisive hands.

Altogether, Jaguar built 54 C-types. That's all it took to achieve sports-car immortality.

CHAPTER IX

# XK 140: IMPROVING A CLASSIC

A curious paradox that troubles thinking auto buffs is the way the conception of "better" changes over time.

If a new model happens to be one of those rare designer bullets that hits all our intellectual and emotional bull's-eyes, we may call it "perfect," a "modern classic" even. "Couldn't be better," we rave.

Then, a year or two down the road, we begin admitting that one or two areas, very minor, could just possibly stand to be a little better. Perhaps we start envisioning different wheels, seats, suspension parts. After four or five years, our list of potential improvements has outrun the design's ability to accommodate them. The car just can't embrace the newest technology. Meantime,

maybe *we've* become taken with newer models.

"What an obsolete old crock," we finally exclaim. "What's wrong with that company? They should have replaced that thing years ago!"

Of course. We ride the tides of our times. We use our cars in our world, and our world keeps moving. Tires get better. Brakes improve. Suspension, chassis, body, transmission, engine concepts advance. So do roads. And laws change. Fuels, too. We ourselves evolve in our needs and notions. By 10 years of age, a car really seems old in more ways than one.

Ah, but in another 10 or 20 years that same car may again be hailed a "classic," even if it doesn't merit that oft-misused term. And the more "classic" it is, the more removed from the present, the more nostalgically attractive it seems—and again "perfect" somehow. "They don't build 'em like this anymore," we pridefully gush. "Best thing on the road. Couldn't be better."

By 1954, Jaguar's XK 120 had been known for six years and was familiar for four. Coventry had unleashed slightly more than 12,000 of them. It was still just as beautiful as the very first of its kind in 1948, yet had become a better car in numerous ways. Some people had even begun calling it a "classic." Nevertheless, changing times seemed to demand something more in the way of improvements than Jaguar had mustered so far. They arrived that October in the form of an evolutionary replacement, the XK 140.

At a glance—a pretty good look, in fact—it appeared to be just an XK 120 with more chrome. One immediately noticed the new "wide-stripe" grille. The solid, simple one-piece casting with fewer and thicker bars replaced the original frontispiece with its many slim vanes, which had a weakness for getting bent out of shape, and was expensive to make anyway. Also obvious was the addition of a full-length chrome strip to the hood and trunklid. Completely invisible was the adoption of steel doors except on the roadster, which remained more of a stark lightweight suitable for amateur competition. "Bonnet" and "bootlid" on all three body styles continued to be rendered in aluminum.

Bumpers, which had been sketchy things mounted directly to the body, were deeper, more robust, newly wrapped to the sides for added protection, and attached to the frame for added effectiveness. They increased overall length three inches, but neither this nor their extra bulk was readily apparent. Lighting equipment became more generous with the addition of flashing turn signals (as small circular units outboard just above the front bumper and as integrated elements within enlarged taillamps). There also were brighter headlights, supposedly stemming from Le Mans experience.

But beneath all this, the 140 still was essentially a 120, with the same captivating, if aging, concatenation of catlike curves. Wasn't it?

Well, yes and no. In many ways, the 140 was a substantially different car.

Though both the chassis and the body's basic dimensions and shapes were altered only slightly, Jaguar had gone to a great deal of trouble to give its customers more interior room. Right away, old 120 hands applauded the replacement of the "bullet" horn button with a flat item. After getting in, taller people were happy to find an inch more space above the knees. Jaguar had in fact raised the entire body on its mounts by an inch. Even more welcome to such unfortunates was the extra legroom conferred by pushing the firewall three inches forward, which allowed an approximate doubling of total seat

*The XK 140 bowed in '54 looking much like the XK 120 it replaced. But there were changes beneath the skin and Le Mans wins on its badge.*

travel. At the same time, the entire back wall of the cockpit was redesigned, and the twin six-volt batteries relocated from that area to the engine bay. That made room that could be devoted to small "occasional" seats just ahead of the rear axle, or to a folding top, depending on model. For those who had felt constrained by limited cargo volume, the 140 offered a little more, thanks to a lowered trunk floor.

Structural alterations went even further in the coupe. Its entire instrument panel/windshield assembly was moved forward three inches, and the firewall was bulged forward either side of the engine, all so the two front seats could be stationed farther ahead. Together with moving the back window to the rear by nearly seven inches compared with the XK 120 coupe, the extra cockpit length made the two back seats almost habitable. A thoughtful touch was a provision for relocating the back cushion of one of these kiddie seats to the opposite corner, so a single consenting adult could ride side-saddle. Another was a hinge-down panel into the trunk that allowed long items to be carried.

Coupe doors were now 5½ inches longer, and all this added length called for redrawing the arch of the roofline. That provided more than an inch more headroom, plus larger side and rear windows that gave a much brighter interior—and a better view out of it. If the 140 coupe looked a little pudgier than its 120 predecessor . . . well, perhaps so did some old Jaguar partisans by now.

Moving the firewall those three inches required shifting the engine/transmission package the same distance forward on its chassis rails. That meant the car was no longer tail-heavy, but carried slightly over half its weight on the front tires. Stiffer torsion bars—the parts from the previous Special Equipment option, in fact—were standardized to carry the extra weight. The dynamic result was a re-tune of the chassis' basic handling characteristic to more resolute understeer.

*In standard tune the XK 140's six had 190 bhp, but 210 bhp—the magic one horsepower per cubic inch—was available with the new "C" package. Triple-carb setup is not factory-original.*

*A flat horn button replaced the old "bullet" hub, while rack-and-pinion steering was added. The four-speed manual got a less-frenzied first gear, and optional overdrive and automatic were made available. Pushing the firewall forward by three inches increased leg room and seat travel.*

Also important was the adoption of rack-and-pinion steering, as on the C-type racer introduced three years before. Compared to the superseded recirculating-ball system, this more direct mechanism improved both feel and response at the cost of some additional kickback through the steering wheel. Another cost was an increase in turning circle, from 31 feet to 33 (also the same as the C-type's). A benefit was a happier steering-column position. Now running lower through the engine compartment, it left ample clearance for any future additions to the carburetors, though that may or may not have been foreseen at the time. Jaguar also seized the chance to add a universal joint just ahead of the firewall to give an additional cant to the column. That answered some complaints that the steering wheel was too upright and its rim too close to the thighs.

Jaguar also answered complaints that the dual-exhaust system hung too low. Because the frame's main central crossmember had to be revised to suit the new transmission-mounting point, engineers decided to lead the pipes through holes in the crossmember, one on either side of the centerline—neat.

The main change at the rear was to telescopic shocks, which were considered more durable than the old lever-type. Their upper mounts doubled as strengthening gussets in the chassis crossmember atop the axle, another tidy touch. Also, rear track went up by 1$^3/_8$ inches, to 51$^3/_8$.

Despite Jaguar's success with disc brakes over the past three racing seasons, the heavier, faster XK 140 retained the 120's fade-prone drums. At least they could now be fitted with competition linings at extra cost.

The 140 also retained the 120's four-speed gearbox. But soon after production began, its three lower ratios were moved closer to top. The big advantage of this numerical lowering was to eliminate the frenzied gearing of first, which most 120 owners had found useless. That cog remained unsynchronized, however.

The rear axle brand was now Salisbury, as on the C-type, a change actually carried over from late XK 120s. It brought a slightly longer-striding final drive of 3.54:1, up from 3.64. The 140 marked another first for Jaguar's sports car in offering optional overdrive, the same Laycock de Normanville unit already in service on the Mark VII sedan. Bolted to the back of the gearbox, it engaged via a facia toggle to provide the same effect as the overdrive top gear in one of today's five-speed manuals. So equipped, rear-axle ratio was a much lower 4.09:1, but switching in the OD gave an effective final drive of 3.19:1, which made for very relaxed cruising. A later transmission innovation for the 140 was an optional automatic, the American-made Borg-Warner three-speed also in use on the Mark VII.

With all these big differences, and many detail ones, Jaguar fans must have been relieved to see that the wonderful twin-cam six had been left pretty much alone for the 140. There were changes to be sure, but like the car itself, they were of the considered, common-sense variety. The most important were a switch from cast iron to stronger steel for the main bearing caps, and a smaller sump made of pressed steel instead of cast aluminum, this to reduce the chance of road damage. The old overheating problem was addressed by moving the radiator forward at the bottom, the resulting slant creating more core area for cooling, and adopting an eight-blade fan shrouded in a sheetmetal duct.

Standard engine tune was now similar to that of the previous SE option, called "Modified" (M) in the U.S. This comprised cams giving $^{3}/_{8}$-inch lift, 8.0:1 compression, and the pair of familiar 1$^{3}/_{4}$-inch SU carburetors. Despite only a single standard exhaust pipe as before, outputs were satisfyingly healthier: 190 horsepower at 5500 rpm and 210 pounds/feet of torque peaking at a modest 2500 rpm.

The SE/M option was still around, but with its uprated engine standard, the 140 package was little more

*A new grille of thicker vertical bars, a chrome strip running the length of the car's spine, and heavier bumpers helped distinguish the XK 140. Also, steel was adopted for the doors of the Fixed Head Coupe and Drop Head Coupe.*

than wire wheels and a pair of foglamps. But buyers could now combine this with a new extra-cost C package, offering still higher performance courtesy of the Le Mans-proven C-type cylinder head. Identified by the letter "C" cast right into it, this brought larger exhaust valves and larger ports both sides, as well as dual tailpipes. Special pains were taken during assembly to match ports with manifolds. All this boosted the 210-cubic-inch XK six to a smashing 210 bhp at 5750 rpm—"1 h.p. per cu. in." a full year before Chevrolet ballyhooed that achievement with its fuel-injected 283 V-8. The absolute torque value of the C-head engine was only 213, but the peak was way up at 4000 rpm. XK 140MCs also received competition linings on brake shoes and, at least sometimes, roadworthy racing tires.

For owners who really wanted to uphold the racing tradition, the factory was happy to offer an even stronger motor, with 9.0:1 pistons, two-inch carbs, special bearings, and a solid-center clutch. It was cataloged on all three body styles, but when the lucky car was a roadster, the old aeroscreens and bucket seats could still be ordered. However, not much else in the way of performance equipment seems to have been offered—none of the special springs, bars, shocks, bushings, brakes, wheels, etc. that would have given the impression that the factory expected its clients to race their 140s seriously. Perhaps Jaguar considered it enough to provide them *real* race cars in the C-type and the D-type that was just now replacing it.

What about that designation, XK 140? What happened to 130? Nobody seems to have explained that at the time, or to have bothered about it. After all, the XK 120 had proven able to beat 130 mph, so there was no point in using that number for a supposedly better model. And in fact, the 120 had been pushed up above the 140-mph frontier. In May 1953, one had been stripped down and taken back to Jabbeke in Belgium, where in official timed runs it had gone 141.8 mph. That was obviously justification

*Docile around town, it could be a dervish on the open road. The fastest of the XK 140s would run 0-60 mph in 8.4 seconds and top out at 129 mph. As on the XK 120, the tachometer read counterclockwise, but now it was surrounded by burled wood rather than by leather.*

enough for hanging 140 on the revised model.

Whatever the figure, the 140 by all accounts was received with enthusiasm. *Autosport*'s technical editor, John Bolster, called it "a great improvement in every important respect" over the 120. Testing an MC coupe, which overall impressed him as "a delightful piece of work," this tallish man praised the roomier accommodation and the "far better" driving position. He even described the "delightful little gear lever" as "ideally placed," an opinion not always shared by shorter colleagues who had to slide the seat farther forward.

Moving off, Bolster discovered 210 horsepower was enough to provoke wheelspin "on any road surface." But he got this heaviest of 140s (he quoted 26.5 hundredweight, or 2968 pounds) to 60 mph in 10 seconds flat despite having to change gears twice on the way, at 30 and 52. He needed another shift at 76 to finish the quarter-mile, doing the job in 16.8 at 80. This performance he termed "immense." He also noted that the overdrive ratio was so tall that his test coupe's corrected maximum speed was only 121.6 mph at an engine rpm of 4800, nearly 1000 short of the power peak. But that didn't matter, he said, because the real value of the Jaguar was its ability to get quickly to 100 and then maintain it as a cruising speed "with only a whiff of throttle." The 140 was "the most effortless car imaginable."

As for handling, Bolster did admit that the leaf-sprung rear axle—which was beginning to seem old-fashioned even in 1955—could upset the car's balance on bumpy corners. For the most part, though, he found "a wonderful feeling of complete control . . . one of those very rare cars that seems to help the over-impetuous driver out of his difficulties. The rack and pinion steering gives beautifully precise control, and one can fling the 26 cwt. car around like a 1½-litre machine." He characterized the suspension as firm, so that "all the 'float' of the 120 has gone." And once he got the racing linings properly bedded in, the brakes just did not fade in repeated trials

*The XK 120 was a strict two-seater, but Jaguar gave the XK 140 coupe and convertible a pair of small rear seats. Additional structural changes to the coupe made its aft compartment almost habitable, provided more head room in an airier cockpit, and improved visibility.*

from 120 mph. "Driven hard," gas consumption worked out to 14.6 mpg (converted to U.S. gallons).

"If the XK 140 is an extremely fast car," Bolster continued, "it is also a very docile one. It has absolutely perfect traffic manners, and entry and exit is easy, even for a lady in evening dress. The excellent all-round visibility and compact dimensions are also appreciated in town, just as they are on the open road. Above all, though, this is an ideal long-distance car, and would be sheer heaven for Continental touring."

He was right. The self-same test vehicle was taken to Belgium and France immediately afterward by *The Autocar*, and after 588 fast miles in sports-car heaven, by and large the comments were similar. "The readiness with which 100 m.p.h. can be reached . . . is the car's outstanding attribute." Once there and with the OD switch on, "the car seemed to don the proverbial seven-league boots, and the tall popular [*sic*] trees lining the long straight stretches of road appeared like a giant fence rushing by."

All this speed, yet impressive silence too. *The Autocar* noted some engine induction roar, but no audible exhaust noise, and little from the wind. Normal conversation was easy. This virtue plus a roomy interior, a firm but comfortable ride, stable and reassuring handling devoid of the XK 120's tendency to oversteer, and the very powerful but amazingly flexible, fuss-free engine added up to a vast departure from the traditional sports-car driving and touring experience.

Not surprisingly, we find some performance discrepancies between these two tests. At first glance, one might guess that Bolster's runs had freed the engine a little, for the rival testers reached 4900 rpm in OD, which they said was equal to 129.25 mph. But the math doesn't work out; another 100 rpm should only have added 2.5 mph to Bolster's 121.6. *The Autocar*'s carefully measured top speed really was faster, though, because the Jaguar had been reshod with racing tires. These had both a larger diameter (rolling radius) and, presumably, more resistance to the power-robbing "standing wave" phenomenon that afflicted ordinary tires of the day at high speed.

Perhaps the larger tires also robbed some acceleration, for *The Autocar*'s 0-60 time was a sluggish 11 seconds, the quarter-mile's an equally lackluster 17.4. But then, *Autocar* always tested with two people aboard, whereas Bolster drove solo. Despite this, its fuel economy worked out to a more palatable 18 mpg U.S.

Weight as given by the second magazine was easier to credit, being 28 cwt or 3136 pounds with five Imperial gallons (40 pounds) of "first grade" petrol aboard. Front/rear distribution was 50.3/49.7.

Jaguar fans anxious about possibly losing the more sporting aspects of the old XK 120 may have been reassured by *Road & Track*'s June 1955 report on an XK 140MC roadster. Zero-to-60 time, the most important automotive statistic on the American continent, was a sparkling 8.4 seconds. Whew.

Crunching other numbers in the *R&T* data panel gives a more rounded picture. To begin with, theirs was an early 140, and thus had the old wide-ratio gearbox without overdrive. That meant the overall gearing was such that only one shift, at 39 mph (5800 rpm), was necessary on the way to 60. Another change to third gear at 68 took the car through the end of the standing-quarter in 16.6 seconds at about 82 mph. Very little better than *Autosport*'s coupe.

And *R&T*'s observed top speed was but 120.3 at 5200 rpm, 550 rpm short of peak power. Of course, an open-cockpit car with the normal windshield could be expected to be "draggier" than a coupe. Yet back in 1949, private Jaguar tests had established that a 160-bhp XK 120 of the same configuration, and without bellypan, could do exactly 120 mph. Hmm. The 140MC's extra 50 horses must have really been laboring to pull the bigger

*Its lines were timeless, but sluggish cold starts, heavy steering at slow speeds and kickback over bumps, firm clutch pedal, long 1-2 gearlever throw, and obscured gauges were some complaints.*

*Coventry shifted the XK 140's engine and transmission forward, eliminating its predecessor's tail heaviness. Roadsters like this one tipped the scales at about 3135 pounds; coupes were around 100 pounds heavier.*

bumpers, added foglights, wire wheels (without rear spats), and extra weight all of 0.3 mph faster.

How much extra weight? It seems likely that by "curb weight," *R&T* meant that the 16.8-gal. gas tank was full. If so, then subtracting 6.2 pounds for each gallon from the stated curb weight of 3135 leaves an empty XK 140MC roadster at 3031. That's only 65 pounds less than *Autocar*'s figure for an empty coupe, but perhaps a couple of hundred more than an XK 120 roadster.

As had their British colleagues, whose opinions they so often shared, the American journalists decided ultimate speed was not as significant as other elements of the Jaguar equation. "In our opinion," *R&T* declared, "the 'standard of the world' has been, and still is, the Jaguar—in the sports car category." There was enough performance "to cruise at any speed up to 100 mph, for as long as road conditions permit." Above 75, though, wind noise became too loud for normal conversation. Compared to the 120, which the editors admitted they hadn't driven for nearly four years, *R&T* thought the 140 was a little easier to drive in traffic, had a nicer gearchange (though it still didn't like to shift quickly), and a quieter and smoother engine unless full power was used. New mufflers made the exhaust "nearly dead-silent, in marked contrast to the crackle of the earlier M-type mufflers."

They also approved of the bigger trunk space and the additional legroom, but remarked that, on their left-hand-drive car, the pedals were offset to port a bit awkwardly. The new steering system was deemed "without fault," although one driver thought effort excessive in hard corners. (Perhaps in response, Jaguar later reduced caster angle to lighten things up.) No remarks were made on actual cornering behavior, beyond opining that the "considerable roll" was "a reliable indication of safe cornering speed and such a car as the Jaguar reduces the 'driver skill required'; it is in fact one of the easiest and safest automobiles to drive being built today."

At a 1955 list price of $3745, including all the M

and C hardware, the XK 140 roadster was a remarkable value—actually several hundred dollars cheaper than originally charged for the XK 120. Yet it seemed to be high in quality. As *Road & Track* observed: "With performance per dollar excelled by no other car, the nicer details of finish and fittings on the XK roadster come as a pleasant bonus feature. The quality of finish is immediately apparent on the outside, but a look under the hood shows attention to detail that is in marked contrast to that found under a domestic product."

Ownership experience, *R&T* continued, had apparently been good, "a remarkable record of owner satisfaction which stems from good service." In later years, the magazine would have quite a different view on this point.

Jaguar played to a tougher house when it handed over an MC roadster to *Sports Cars Illustrated* for its issue of March 1957. Maybe it was the additional two years of experience, but the always-thorough Karl Ludvigsen found quite a few things to list against the 140. These included "sluggish starting on cold days," torque delivery showing "a strange flat spot between 2000 and 3000 rpm," clutch actuation that was unpleasant both because of heaviness and the pedal's still-acute static angle, overly wide accelerator/brake-pedal spacing, heavy brake feel with the racing linings, brake cooling vents that could admit rain water, a too-long 1-2 gearlever throw, slow synchromesh, heavy steering at low speeds, disproportionately extreme steering-wheel oscillations set up by the rack-and-pinion's inherent kickback, and obscuring of some switches and gauges. Oh, and a top that was hard to put up, hard to keep together at speed, hard to see out of, and hard to stay dry in.

So did Ludvigsen hate the 140? Not at all. His list of likes was far longer and stronger. "This latest revision of a time-tested machine is notably improved in the handling and braking departments," he reported. "This together with its smooth and surging power make it a delight to drive at high speeds over long distances on fast, winding roads. If used in town it can be difficult to the point of being tiring, but this is not its purpose in life." The car arrived at its "area of greatest competence" once third gear was engaged, with the 3-4 shift movement being "short and neat." He loved the smooth, solid-feel-

*Ill at ease in town, the XK 140 lived for "high speeds over long distances on fast, winding roads." When production ended at 9000 in '57 the design was 10 years old, but to many it still was the standard of the sports-car world.*

ing big engine and its "instant response" above 3000 rpm, when "the nose comes up and the roadster begins a relentless rush forward. Third gear with its maximum around 100 [mph] is particularly nice and just right for positioning and power out of open back road bends."

This unabashed sensualist also enjoyed the "familiar sophisticated Jaguar exhaust purr," the "machine-like whir" of the dual cams and their drive chains, and even a "pleasing mechanical whine" from the transmission—"part of the car's appeal." So was the appearance of the C-head engine: "an unalloyed joy with its red and polished aluminum finish, and the enamelled exhaust manifolds [which] recall the classic era."

So in 140 form, the XK was as much a modern classic as ever, this second iteration merely putting more emphasis on the "modern" part. By the time the 140 departed in early 1957 with just more than 9000 made, nearly a decade had passed since Sir William's great hunting cat first stalked the world's sports-car *aficionados*. But the fabulous XK still had the power to excite and delight like no other car. A good thing, that, as it still had a lot of stalking to do.

*Prod the twin-cam six and acceleration was "immense"; 100 mph was maintained "with only a whiff of throttle," and the XK 140 was "the most effortless car imaginable."*

CHAPTER X

# D-TYPE: FOR THE LOVE OF RACING

Every racing machine is a labor of love, but few race cars have been so intensely loved as the D-type Jaguar. Much of that arises from sheer physical beauty. Seldom has road-racing science blended so perfectly with road-racing art. But much of the adoration also stems from what this car accomplished for its company and for its nation. With three straight victories at Le Mans—in 1955, 1956 and 1957—the D-type clearly demonstrated Jaguar's mastery of its chosen subject: The very high technology involved in building very-high-performance automobiles.

Curiously, it was a demonstration the company really didn't have to make. For by all objective measures, Jaguar didn't need to build the D-type.

On the morning of Monday, June 15, 1953, Jaguar Cars opened for business as the automaker that had just won the Le Mans 24-hour sports-car endurance race for the second time in three serious attempts. This victory by the sleek, dark-green, disc-braked C-type sports-racers had been particularly convincing. The publicity was intense, favorable, and reached world-wide. Jaguar's entire workforce must have arrived that morning with their backs just a little straighter, knowing that their marque had a secure place in both its owners' estimations and automotive history. Even the Queen had sent congratulations.

Some automakers might well have stopped there. What more could possibly be proved by going back in 1954? Nothing, surely, that was worth the risk of a loss undoing all the good achieved so far. Withdrawal could have been a simple matter of a press release mumbling something about the need to concentrate on the passenger-car range and applying to it the technical lessons learned in racing for the good of the company's loyal customers.

But not Jaguar, not now. The firm was on too much of a roll, had too many important and interesting problems to solve, and was simply having too much fun. Why, the very name had come to mean high performance. After all, Jaguar had just won two out of three. Quit racing? Impossible.

And in fact, work toward a new competition Jaguar was already well along, with aerodynamics a prime field of investigation. While the low-drag "droop-snoot" had been a failure in 1952, prompting the original C-type body to be readopted for 1953, chief engineer William Heynes and aerodynamicist Malcolm Sayer had not lost interest in streamlining. Nor could they afford to. Le Mans was the centerpiece of Jaguar's racing program, and it placed a premium on top speed.

On the 8.38-mile circuit then in use, the straightaway known as Mulsanne to the English (*Les Hunaudieres* to the French) was a single blast of wide-open throttle 3½ miles long. The value of "good aero" there was both obvious and substantial. If a car could go just a few miles per hour faster, the time it spent on this one straightaway would be cut by whole seconds. That would be a sizable advantage over rival cars, which would have to really scratch to make up time through the other 58 percent of each lap.

"High speed" in 1953 meant something above 150 mph. Before it overheated, the "droop-snoot" had shown itself capable of about 152 mph along the Mulsanne Straight, some eight mph better than the 1951 car of almost identical horsepower. In 1953, it had taken substantially more muscle to push the readopted original-body C to a best of 148.8 mph. Streamlining was the way to go.

Besides, speed sold cars. Jaguar had been speed-conscious since the very launch of the XK 120, whose model number promised that many miles per hour. The first public demonstration of the sports-car's performance on that closed superhighway near Jabbeke in Belgium in 1949 was devoted to proving the promise, and the stripped-down 120 actually went 132-plus on that occasion.

That was with the original 160-horsepower twin-cam XK six. Four years later, in April 1953, a 180-bhp XK 120 roadster equipped with bellypan and small windscreen demonstrated an impressive 141.8 mph at Jabbeke. The same car returned the following October, albeit much more highly modified, and test driver Norman Dewis vaulted it to the astonishing velocity of 172.4 mph. This astounded even Jaguar's own people; privately, they admitted they'd have been happy with 155.

But on hand that same day was another Jaguar that went 178.3 mph. This one was a virtual streamliner, a very small, very smooth egg-shaped car whose single driver's seat was capped by a tiny plastic bubble taken from a sailplane. It was even more advanced under its

*Epitome of the road car as racer, Jaguar's voluptuous D-type carried Coventry to three consecutive Le Mans wins in the fifties.* Far left: *The long-nose version took Mike Hawthorn to victory over Mercedes' Juan Manuel Fangio in the tragic '55 race.* Left: *This short-nose car won Sebring the same year.*

eggshell, for the chassis was not a steel-tube space-frame, as on the C-type, but a stressed-skin structure, or monocoque largely made of magnesium sheetmetal.

Though never formally announced as such, there were enough rumors and, subsequently, official evidence, to be sure this car had originally been intended for the 1953 Le Mans race. But it is just as clear that the sour experience of 1952, when the "droop-snoot" body was rushed to race without sufficient testing, led to second thoughts about this new experimental design. It thus stayed at the factory, deemed either unready or in some other way unsuitable for the 1953 event. But as often happens with such cars, it was pressed into service as a research vehicle and covered many thousands of miles on various test tracks before finally being revealed at Jabbeke in streamliner form. We can also be sure that it greatly helped Heynes and company in designing what did appear as their next Le Mans car.

One other interesting prototype was secretly built and tested in this period, though it was more of an unofficial sideline and led to less. William Lyons, whose company was basically founded on his sense of style, was always experimenting in full-scale with various body ideas. Possibly to help himself come to grips with true streamlining, a big change from the more traditional forms he'd long since mastered, he spent much of 1953 playing with a vehicle along the lines of a racing sports car like the C-type.

Looking very aerodynamic, it had a low, rounded nose and a sleek, unbroken torpedo profile leading to a tapering tail. All four wheel openings were enclosed with spats ("valences" in Jaguar parlance), and the driver sat behind a small cowl and curved plastic aeroscreen. So far, so *avant garde*. And indeed, tests carried out by Dewis that September indicated drag was about 15 percent less than the contemporary C-type's. By that stage, Lyons seemed to have been thinking of setting some sort of speed record with his hobby car, for one of the tests involved filling the rear axle with soap as a low-drag lubricant. The project was soon abandoned.

The prototype was an interesting sideline, but unfortunately the overall visual impression was of a very ungainly contraption, with tall and wide fender lines giving a thick-in-the-middle look, and a seat mounted so high that the driver stuck out comically. Around the factory, Lyons' strange beast was known variously as "Bronco" or "The Brontosaurus." It was probably the ugliest thing he ever did.

No, the correct path toward the future was shown by Heynes' speedy-looking magnesium egg. Out of it in the spring of 1954 emerged the immortal D-type.

The letter, of course, implied a simple follow-on to the C-type, whose name indicated the "competition" version of the XK 120. In a way characteristic of the often-shifting logic behind Jaguar model nomenclature, the new racer became known as the D-type even before its chassis

*Jaguar's second significant visit to Jabbeke, in '53, saw its 172.4-mph modified XK 120* (left in photo below) *in the company of a D-type forerunner* (right in photo). *This prototype pioneered Jaguar's D-type monocoque racing design and topped 178 mph.* Opposite bottom: *It's shown without its plastic bubble top.*

numbers were given the suggestive prefix XKD. But the "D" might easily have stood for "departure," because in many ways, that's just what the D-type was.

Certainly, basic layout and much hardware were carried over from the C. An XK six still resided in front and drove through a four-speed gearbox to a live rear axle.. Front suspension remained independent, and there were all-round torsion bars and telescopic dampers. Front radiator, rear rubber-bag fuel tanks and spare tire, rack-and-pinion steering, disc brakes and lift-up nosepiece were all familiar, well-tested competition concepts.

But the engine had a dry-sump lubrication system, adopted partly to cure the oil-surge problems sometimes experienced with C-types and partly to reduce the engine's height. Also equipped with a smaller flywheel and three-plate clutch, the revised powerplant mounted nearly three inches lower, which not only allowed a lower hoodline for minimizing drag but also lowered the center of gravity for consequently improved stability. To make room for the trio of Weber carburetors in a physically smaller car, the entire engine was tilted eight degrees to port.

Output was now 250 bhp at 6000 rpm, according to documentation released on the car's announcement, though subsequent information downgraded that by 5 or 10 horses and 200-300 rpm. But more powerful the D-engine definitely was, thanks to 9.0:1 pistons, new cam timing, larger intake valves, and a more aggressively tuned exhaust system. Torque came to 242 pounds/feet at 4000 rpm.

Continuing the departures was a completely new gearbox with bottom-gear synchromesh for the first time. Rear suspension was modified to provide more positive axle location, and much of the front suspension hardware was new, with that elegant, satisfying look of purpose-built forgings. The wheels, no longer the familiar old wire-spoke affairs, were a stronger, lighter disc design by Dunlop.

Strong and light aptly described the D-type's new chassis. Its middle part, where the driver and theoretical passenger sat, was a "tub," a monocoque structure made of folded, riveted and arc-welded light alloy. Jaguar's monocoque was not quite the first-ever application of this aircraft construction technique to auto racing, but it was the first of significance.

Adding strength to the central spine of the monocoque and completing the chassis forward of the firewall was a space-frame mainly made up from square-section tubes. These were made of aluminum and welded directly to the tub. Gracefully shrink-wrapped around all this lightness was a new bodyshell, the result of much wind-tunnel research by Malcolm Sayer.

Dimensionally, the D was distinctly more diminutive than the C. Its wheelbase was more than five inches shorter, at 90.6. Its front track, at 50 inches, was one inch narrower, and the rear was three narrower, at 48.

*Wind tunnel testing of scale models like the one at right resulted in an aero bodyshell over a structure that combined a lightweight space-frame with advanced aircraft construction. The finished D-type was wider and lower than the C-type and at 1930 pounds, was 10 percent lighter.*

Right: *Not all of William Lyons' creations had the grace of his XK sports cars. This cartoonish experimental streamliner was tested in secret during 1953.*

Turning circle was a foot less, at 32. Nose to tail, the D-type body initially measured 154 inches (versus 157), though later it would grow a little longer. From the start the D was a trifle wider, its aerodynamic fenders swelling to 65.4 inches, though that was just 1.9 inches up on the C's maximum width, which was established by its protruding wheel nuts.

But the new car was lower through the body and, in original form, had a windshield for the driver alone, so frontal area was only 12.5 square feet, a useful 9.4-percent reduction from the C's 13.8 square feet. Drag, which Jaguar measured as horsepower required to maintain 100 mph, was down more than 28 percent (38 bhp versus 53). Even against the "droop snoot" C, the D-type was cleaner by 4 bhp, or almost 10 percent. The extra slipperiness was apparent on the Mulsanne Straight, where the 1954 D clocked nearly 174 mph, some 14 percent faster than the unstable 1952 car.

As always, numbers for weight are suspect, but a D-type figure used by the factory itself was 1930 pounds, about 10 percent less than the C-type. Distribution was 1000 front/930 rear, or 51.8/48.2 percent. This was in "dry" condition, of course. Aerodynamic testing figures published by Jaguar's Bob Berry in *Jaguar: Motor Racing and the Manufacturer* show a presumably race-condition weight distribution even more nose-heavy, 53.5/46.5. Yet at a speed not specified, but probably 100 mph, this reduced to 51/49. That meant the new car was not only faster than the ill-fated "droop-snoot," it was far more stable, actually gaining effective weight on the back end. The whole car may have been lifting, but at least the tail wasn't lifting more than the nose. And once a prominent fin was added to the headrest, D-type drivers reported they could scream along at 170-plus with their hands off the steering wheel. It was restful, they said.

All these data may seem tedious when presented in raw form, but the connect-the-numbers picture they paint is of a very thorough, dedicated racing department doing its job with scientific seriousness and painstaking precision. Jaguar had gone way beyond the "backyard-special" stage.

Racing, however, had not yet progressed to the "18-wheel transporter" stage in 1954. So in time-honored fashion, the first D-type, still unpainted, was driven over to France on public thoroughfares for springtime testing at Le Mans. The roads making up the circuit had been closed for a rally, and the time allotted was short, but the new Jaguar managed to avoid the officials with their angry flags long enough to beat the 1953 lap record, set by a Ferrari, by five full seconds. Of course, Ferrari had not been idle over the winter, and its 1954 entry arrived with 4.9 liters of V-12 power.

Le Mans that year was plagued by heavy downpours. The three new 3.4-liter Jaguars ran well, better than the cars from Aston Martin, Cunningham, Gordini, Lagonda, and Talbot. Stirling Moss even out-ran the best

*The real fruit of Jaguar's 1953 development work was the '54 D-type prototype* (top, both pages). *It was designed specifically to win Le Mans, but also could be driven on the road.*

Ferrari to take the lead and set his speed record down the Straight—this in the evening, no less, and despite a bout of rain. But then all three D-types started misfiring and had to make long unscheduled pit stops to have their fuel systems cleaned. In its scientific zeal, Jaguar had chosen paper-element filters that were *too* efficient: They became clogged with dust floating in the gas supplied by the race organizers.

With the offending filters torn off, the Jaguars got back up to speed, but this was not going to be their year. Moss lost his brakes at the end of long Mulsanne and retired on the spot. A second car suffered gearbox trouble, then engine failure. The third car, driven by 1953 winners Duncan Hamilton and Tony Rolt, took an off-road excursion that caused some damage and another delay in the pits. But this sole remaining D-type was able to struggle onward, holding second to the Ferrari. A flash of hope came near the end, when the Ferrari balked at restarting during a pit stop, leaving the Jaguar to swish by in the rain and into the same lap. But the Ferrari finally fired and roared back onto the track, just 97 seconds to the good.

The British team tried its hardest, to the point that Hamilton was getting wheelspin in top gear at top speed on the straight in the pouring rain, but the big Italian managed to prevail. The D-type finished 105 seconds behind it after 24 hurly-burly hours. A privately entered C-type placed fourth. The D-type contested two other races in its first year; Rheims was a win, Dundrod a loss.

Over the winter, Jaguar built new D-types for the 1955 season. While the basic design was the same, there were important improvements and refinements.

Whereas the 1954 car's central tub and forward space frame had been integral, the two elements were now separate and attached by bolts. This was mainly to make crash repair easier. To make manufacture easier, the front frame was made out of steel tubes, which could be bronze-welded together by journeymen, rather than the argon-arc specialists the aluminum had required. Yet the steel frame was lighter.

Engine power was up, thanks to a new cylinder-head casting. Both inlet and exhaust valves were now larger, and to get the latter to fit, exhaust-valve angle had to be changed from 35 to 40 degrees from the vertical. The inlet angle remained 35, so the included angle was now 75 degrees, but these were forever known as the "35/40 heads." Camshaft lift was up an additional 1/16 inch, and cam timing was yet another stage more radical. With retuned intake and exhaust manifolding, output was now 270 horsepower over a nice, broad spread of 5500-6000 rpm. To help nail this down out of slow corners, Jaguar installed ZF limited-slip differentials for the first time.

On the outside, the 1955 factory team car looked a little leaner and meaner because the radiator inlet was brought forward 7.5 inches, the better to pierce the air. Brake cooling inlets were added, too. The driver's windshield got more of a wraparound, and the fin was extended more toward the tail. Frontal area was up slightly, to 12.8 square feet, but other body refinements combined to lessen drag a further 16 percent (to 32 bhp at 100 mph) and permitted a slight easing in the front-heaviness (to 52.5/47.5 percent) without any change in the at-speed values.

Hindsight is of no more value in racing than in any other part of life, but it would have been better had Jaguar never gone to Le Mans in 1955. A tragic foreshadowing of what was to come occurred when William Lyons' only son, John, was killed in a head-on crash with an American army truck on his way to the track.

At Le Mans for its second year, the D-type Jaguar faced not only Ferrari with its all-new, 4.5-liter six-cylinder cars, but Mercedes-Benz, back in the sports-car fray with a two-seat, 3.0-liter version of its fuel-injected, straight-eight Formula One car. Aston Martin and, this year, Maserati were contenders, too. During the first stage of the race it was a Ferrari in front, but it faded after a while, leaving a fierce battle between the Mercedes of Juan Manuel Fangio, who had already won two of what would be five world driving championships, and Jaguar's Mike Hawthorn, a future champion in his own right.

Against the D-type's disc brakes, the Mercedes 300SLR was deploying an air brake, a large flap behind the cockpit that the driver could hinge up into the airstream to help slow the car for corners. Germany's "silver arrow" was thus able to stay with the green British bullet on the brakes, and it showed superiority through the corners themselves. But the Jaguar had better dig away from the turns and better top-end on the long straight. Effectively equal around the track as a whole, two of the fastest men in international racing were putting on a real Grand Prix. Inspired, Hawthorn set a lap record at an incredible 122.39 mph.

Then he came up to the pits for his first refueling stop. Trying to save time by not braking until the last instant, Hawthorn swerved right toward the pits in front of American driver Lance Macklin's Austin-Healey. The slower car veered left, into the path of a silver Mercedes coming up behind—not Fangio's car but that of teammate Pierre Levegh. The Mercedes launched off the Healey and flew diagonally to the left, over a low embankment into an otherwise unprotected crowd of standing spectators. More than 80 were killed in addition to Levegh. It was the worst accident in the history of motorsports.

All heart was gone from everyone, but, fearing worse confusion if they stopped the race, officials let it continue. The Jaguars lagged, and the surviving Mercedes cars droned on ahead until an executive order was telephoned from Stuttgart to pull out. Hawthorn and co-driver Ivor Bueb soldiered on around to claim a hollow and very somber "victory," one which Jaguar refused to advertise.

Racing, like life, continues. Later that summer, at

*After falling short at Le Mans in '54, the D-type came back with a vengeance in '55. The 3.4's horsepower was up by 20, to 270, and a new nose improved penetration.* This page: *the Mike Hawthorn/Phil Walters '55 Sebring winner.*

Dundrod, Hawthorn and Fangio renewed their titanic struggle. Once again, the Englishman's Jaguar was slightly faster than the Argentinean's Mercedes and set a new lap record. That all came to naught, however, because Moss, also in a Mercedes now, was already ahead, and Hawthorn blew his engine in trying to catch up. At the end of the season, after Fangio had captured his third world Formula One championship, Mercedes announced its withdrawal from all racing.

For the 1956 season, Jaguar again fielded a team of new and improved D-types. These had to meet regulations framed in the aftermath of the Le Mans tragedy aimed at slowing things down, so they had full-width windshields. While that raised frontal area by 13 percent, to 14.5 square feet, Sayer cleverly continued the plastic over the top of the passenger's seat, making a transparent roof that minimized drag on that side. Another new rule reduced fuel-tank capacity from 45 gallons U.S. to 33.5. Apparently to counter this, Jaguar adopted Lucas fuel injection and set it up for better low-end pickup, not more power. But while better fuel efficiency might have been a benefit, it wasn't evident. Economy was likely served just as much by thinner-gauge metal, so the cars were again lighter than the year before. Their handling was improved by adding a sway bar to the rear suspension (a tweak first tried at Dundrod) plus a stiffer one to the front. To save time in pit stops, the disc brakes were given quick-change pads.

Another item was tested but not adopted for Le Mans: a de Dion rear suspension. The D-type had been designed specifically for the French circuit with its ultra-long straight, predominantly medium-speed corners and unusually smooth surface, and Jaguar's traditional live rear axle worked well enough there. On other tracks, though, the D-type sometimes lost ground through wheelspin out of slow turns and wheel hop over bumps. The de Dion system, then common on Grand Prix cars and some of Jaguar's more sophisticated sports-racing rivals, was a halfway step to full independent suspension. It retained the basic beam-axle geometry, so that the driven tires remained squarely on the road, but relieved the wheels of carrying the burden of a differential. In back-to-back tests, it seemed to make the Jaguar a little faster around some tracks. It would probably have shown no advantage at Le Mans, though, and as a still-new, somewhat heavier design, it was left off.

One additional interesting Jaguar project at this time was an experimental chassis tub fabricated at least partly in that still-exotic postwar material, fiberglass. This was kept secret and apparently never tried in a race, although a car built around it did appear once in public, Mike Hawthorn driving it through the streets of Coventry for public-relations purposes. Though nothing came of Jaguar's "plastic fantastic," it was in 1956 that Colin Chapman began developing his all-fiberglass roadgoing Lotus Elite two-seater coupe, which was unveiled in the

*Three months after Sebring, Hawthorn and Ivor Bueb gripped this steering wheel and throttled this glorious engine for 24 hours to win Le Mans. An accident that killed more than 80 people drained all joy from the event, however.*

fall of '57. It would be another five years before the Chaparral from Texas introduced fiberglass chassis structures into racing.

One of Le Mans' slow-the-cars rules for 1956 mandated an upper displacement limit of 2.5 liters for prototypes, which caused a number of manufacturers not to participate that year. The D-types qualified as "production" because quite a few had already been made and sold. They thus had the largest engines, and were comfortably fastest in practice. About the only opposition within reach of the factory Jaguars were the Aston Martins, which had been allowed to keep their 3.0-liter sixes, and some privately entered Ds.

Jaguar's list of opponents should have included itself. On only the second lap, running in the rain, a team driver crashed in the Esses; the resulting melee took out a second factory car. Shortly afterwards, the Hawthorn/Bueb D-type began to misfire, and lost a lot of time in the pits while the problem was traced to a cracked fuel injector pipe.

The day was saved by the leading privateer team, Scotland's Ecurie Ecosse. Though not quite as technically up-to-date as the factory's British-green machines, its single dark-blue D-type was fast enough to pull away from a battle with the Astons. After 24 uneventful but tense hours, Ron Flockhart and Ninian Sanderson brought home another winner for Coventry. The straggling survivor of the factory team wound up sixth.

Jaguar now had some self-examination to do. In the three years since 1953, when giving up racing was unthinkable, the company's situation had changed. Then, it seemed incapable of doing wrong. Then, it had a brand-new car all but built. Then, racing's rules had been simple and straightforward, and Jaguar's magnificent XK engine was well suited to them and also still had ample developmental life ahead of it. None of that was true now.

It looked like the time had come to take a break. And that's all it was supposed to be. When Jaguar decided after its dismal 1956 Le Mans not to race again that year and to stay home during 1957 as well, it was to regroup for a stronger effort in 1958. No one thought of the hiatus as permanent. Behind the scenes, in fact, evolution of the aging but still effective D-type continued on behalf of its owners.

The most important of this development work was increasing displacement of the XK engine. Since the early 1950s, several XK 120 owners had tried boring out to or beyond 3.8 liters. The factory now tackled the job properly, revising the block casting to improve cooling and installing wet cylinder sleeves to prevent the cracking sometimes experienced by the homebuilders. Bore was enlarged four millimeters, to 87 (3.43 inches), which with the existing, long 106-mm stroke brought swept volume to precisely 3780.8 cc (230.7 cubic inches). In ultimate D-type form, this engine would realize 306 bhp.

Although never officially more than a good cus-

*A dry-sump lube system for the D-type's twin-cam six cured the C-type's oil-surge woes and reduced the engine's height. A trio of Weber carbs was used and a new gearbox introduced first-gear synchromesh. Displacement grew to 3.8 liters for '57, yielding up to 306 bhp and nearly 179 mph on the Mulsanne.*

*Leading Jaguar's non-works race effort was Ecurie Ecosse, French for "Team Scotland." The Ecurie Ecosse D-type won Le Mans in '56 and returned the next year to outgun V-12 Ferraris and V-8 Maseratis and win again. Ron Flockhart and Bueb drove the '57 winner* (above). *Note the full-width windshield mandated to slow the cars after 1955's horror.*

tomer, Ecurie Ecosse (French for "Team Scotland") assumed de facto status as the Jaguar factory's competition arm in Europe. Taking up a parallel position in America was the Briggs Cunningham team, which had been forced by tax laws to abandon the business of building its own cars. Both continued campaigning D-types, Cunningham winning numerous events at home and the Scots triumphing once again at Le Mans in 1957.

The latter was a decisive demonstration of development over design. Because the capricious French rule makers had again changed their minds on the matter of engine size, new and much more powerful cars entered the '57 event. Arrayed against the 3.8 D-types was a 3.7-liter Aston Martin six, V-12 Ferraris as big as 4.1, and a pair of fearsome 4.5-liter V-8 Maseratis. As expected, the most powerful of these headed the Jaguars in practice and the first part of the race. But, one by one, the bigger cars ran themselves into the ground, and as early as the end of the third hour, an Ecurie Ecosse D-type was in the lead. And stayed there for the next 21 hours.

At the checkered flag, the Scottish Jaguars were 1-2; D-types from other teams finished third, fourth, and sixth. Five starters, five finishers, and Jaguar's fifth victory in the race it had taken as its own highest challenge.

Not that the old-fashioned, small-engine Jaguar that outlasted the newer cars was exactly slow and plodding. One 3.8 D-type was clocked over a measured kilometer on the Mulsanne Straight at 178.8 mph, the highest speed recorded during the entire race.

The importance of the marque's achievement was expressed that week by Gregor Grant in his editorial for *Autosport*: "Jaguar have now won the endurance classic five times—in 1951, 1953, 1955, 1956 and 1957!—equalling the Bentley achievements of 1924, 1927, 1928, 1929 and 1930, and bettering Alfa Romeo's record of four wins in 1931, 1932, 1933 and 1934. But there is a significant difference; namely, that Bentley and Alfa Romeo, revered names in the sports car world through the years, were also in the most expensive price class, whereas Jaguar are in a far less costly category, selling their cars in many thousands in today's highly competitive medium price markets. This latest achievement in the world's greatest sports cars race will not go unnoticed by the world, nor will the fact that the superb six-cylinder, twin-overhead camshaft engine of the competition D-type is basically the same as that in the ordinary 3½-litre Jaguar saloon . . . Bravo, Ecurie Ecosse!"

The beautiful blue cars with the white nose stripes went on to other wonderful performances, but the D-type's day in the sunshine of Le Mans had passed. Though at least one was entered every year through 1960, the design was unquestionably obsolete after 1957. And Jaguar never would manage a comeback in 1958. Or, in fact, for decades.

The reasons were several. Not least was that dangerously conservative view that the passenger-car range

needed the full attention of the firm's technical talent. In truth, there were several new models coming along just then: the XK 150 sports car; a large-engine version of the small sedan; a new large sedan; and even a streetgoing version of the D-type called XK-SS. But such rationalization leads to stagnation. Jaguar was a performance-car company, and by dropping back from the cutting edge of the art it risked dulling its entire product line. There was also the inevitable loss of the excellent publicity so far enjoyed, all of it "free." All in all, Jaguar had gained so much from eight seasons of direct participation in speed competition, 1949 through 1956, that its failure to continue is difficult to justify even on theoretical grounds.

What arose as a practical excuse was a major fire that broke out in the factory during the night of February 12, 1957. Much was saved, however, and thanks to enormous efforts on the part of the entire staff, plus remarkable generosity from corporate neighbors, Jaguar was back to full production levels within weeks. But the losses included a batch of D-type/XK-SS cars and most of the tooling to make more. Worse, there was now so much pressure to get the damaged parts of the factory rebuilt and going that nobody could be spared from that priority. Unlike the situation at some automakers, Jaguar's racing department always had worked side-by-side with the production-line staff; there was, in fact, no distinction between them. Inevitably, after a year or two, it became harder and harder to get back to racing.

Not that the spark went out. Heynes and his colleagues kept up their interest and, from time to time, laid plans for other Le Mans cars. For example, in the mid-1960s they built the long-secret XJ13, powered by a new V-12 engine mounted amidships, behind the cockpit. Sadly, it never reached a starting grid.

So the beloved "D-Jag" would remain the pinnacle of Jaguar performance for many years. And its image as such was enhanced by one happy fact: The D could be driven almost like a street sports car.

One of the first to reveal that to the breathless enthusiast was John Bolster, writing in *Autosport* early in 1955. It came in his driving impressions of the Duncan Hamilton/Tony Rolt car, the very machine that had harried the great 4.9 Ferrari at Le Mans the previous summer on the D's debut.

"'Take her away and enjoy yourself, boy,'" Bolster quoted the jovial Hamilton." But don't drive at much over 165 m.p.h. because she's got the low cog in at the moment.' With that warning, Bolster pressed the button and the dead-cold engine sprang to life, ticking over evenly and quietly.

"The seat fitted Bolster like a glove," he continued. "At first, the pedals seemed small and close together, an impression that vanished almost at once. The steering wheel was also small; in fact, the whole car seemed incredibly low and tiny for a 3½-litre. One sits right down inside the aerodynamic body, and the wraparound wind-

*A cleverly stowed spare tire* (right) *was a D-type nod toward rules governing production-based racers, while the gas-cap in the tail fin was an aerodynamic solution. Its first run at Le Mans in June '54 was unsuccessful* (below), *but a win at Rheims the next month* (left) *sparked a glittering career.*

screen affords such protection that I did not wear an overcoat on that January morning.

"Like all Jaguars," he went on, "the D-Type has a wonderfully smooth engine. It has, in fact, perfect traffic manners, and can be used for shopping without any thought of its potential performance. There is a passenger's seat, but this is normally covered in the interest of streamlining. The steering is light and responsive, the ultra-close-ratio gearbox could not be easier to handle, and the 7 ft. 6 in. wheelbase and 4 ft. 2 in. track add up to a small, nippy vehicle for England's crowded roads."

Comparing the D to the C, Bolster said that the older car's "rear end skittishness" was gone. Despite greater power, wheelspin in the new car was only a problem in bottom gear. The steering he found to be "light and accurate," and he felt less "busy at the wheel" compared to the C. "It feels a much smaller and lighter car than its predecessor, as indeed it is . . ." He went on to describe the ride as "fairly firm" at low speeds, but that it smoothed out at the racer's natural cruising gait. The famous disc brakes, he discovered, were as controllable as they were powerful.

"The acceleration is deceptive, because it is just one smooth rush. The exhaust note is by no means obtrusive, and the once-magic century [100 mph] has come and gone before one engages top gear with the gentle pressure of thumb and finger. Then, road conditions alone determine the speed, and mere words cannot describe the sheer ease of the whole performance."

On a cautionary note, Bolster added, "I do not approve of the possession of very fast cars by inexperienced drivers, but I feel that this is one of the easiest of the real flyers to handle. Unlike some of the latest speed models, it does allow some margin for error, and there is nothing tricky about it. Obviously, its full potentialities on a road circuit can only be extracted by the higher echelon of racing drivers, but one's nearest and dearest could drive it through the West End [of London] without demur."

A car so *demure* that, according to a May 1956 *Road & Track* report, it could rush to 60 in 4.7 seconds and through the quarter-mile in 13.7. Bolster apologized for having had no chance to instrument his similar car, but said no figures could convey "the indefinable feeling of quality that this machine imparts. I am in the lucky position to sample many successful competition cars. Although such vehicles always show high performance, it is frequently accompanied by roughness and intractability, plus some odd rattles and the drumming of body panels. The Jaguar, on the other hand, gives that same air of breeding which the XK coupe possesses. It is, indeed, a new conception in sports-racing cars."

This famous article, oft quoted and long remembered, went down in Jaguar lore as the very definition of ideal sports-car enjoyment. Only 87 D-types were built, and it was very hard to wait for Jaguar to put what it had learned from them into its next generation of road cars.

CHAPTER XI

# XK 150: TAKING CARE OF BUSINESS

Fine art is safe in history. Commercial art has to adapt to the changing needs of its clientele.

The Jaguar XK 120 was a superb piece of automotive art that will forever stand as a milestone. As a product, it stood tall in the marketplace for six years. With the evolutionary XK 140, some aesthetic purity was compromised for commercial values. Significantly, its time ran out after only two-and-a-half years.

Though the successor XK 150 lasted a little longer, it was even more market-oriented. As the final variation

on the stunning 120/140 theme, it directly replied to the likes of the BMW 503 and Mercedes 300SL, which offered similar performance and roadability but better catered to growing buyer demands, especially in America, for refinement and comfort.

William Lyons, a salesman first and foremost, was well aware of how fickle the public can be. His primary intent from the very first 120 had been a "Super Sports" model that would showcase his new XK engine, and he only foresaw building 200 of them. Though orders running into the thousands must have surprised him, he understood that, while people may be attracted by performance, what they often bought was practicality and pizzazz (though Sir William himself is unlikely to have used such a word). Any study of the Jaguars actually purchased proved it. For every hard-core, wind-in-the-hair sports motorist willing to pay for light weight, sharp handling, and raw, rorty power, there were hundreds whose checkbooks opened only for sweet civility and svelte, stylish sophistication.

Accordingly, Lyons spent quite a lot of time and effort working up a total replacement for his original sports-car hit. At least two mockups were prepared for his secret inspection, both reflecting his new interest in aerodynamic body lines. The more finished-looking of these static prototypes, a convertible with a clip-on hardtop, retained something of the XK look about the front end, with a tall, narrow grille and headlights mounted halfway out to the sides. But those headlights were rather Porsche-like, being submerged at an aerodynamic slant into the rounded nose without any nacelles. There was an almost complete absence of the 120/140's sexy waistline dip. The interior obviously gave plenty of elbow room, something the market had begun to notice was missing in the existing coupe, and the windshield was a single piece of curved glass. The whole car appeared a little lower and wider when photographed next to an XK 140 coupe.

Why was this concept abandoned? Perhaps because, overall, the mockup's look was a bit generic, even sedanish. It just didn't have the 120/140's strong personality. No zing. Lyons probably could have fixed that, but there's no doubt that his firm's personnel and resources were already stretched on other projects. Nor was there any indication of the basic 120/140 style wearing thin. Lyons was always sensitive to the "faces" of his cars, and the tried-and-true XK countenance was taking on a handsome cragginess with advancing age. No, a simple revamping of the existing car must have seemed the best idea at the time.

Hence the XK 150. It entered the arena in May 1957 as an interesting mix of the traditional with useful modern touches.

Its chassis, predictably, was the same as the 140's, with the same torsion-bar front suspension and leaf-sprung rear axle, the same fine rack-and-pinion steering, the same choice of disc or wire wheels in the same sizes, and the same basic dimensions. With minor changes, engine and transmission options were also as on the 140. But at long last, Jaguar's racing experience with disc brakes was applied to benefiting the street-car buyer. And not only at the front end, the cheaper solution being advanced by other automakers. The XK 150 had Dunlop's

*Jaguar's final variation on the 120/140 theme was the XK 150. It debuted in May 1957 as a revamp of the existing car. Its styling was stodgier and at first there was no power increase. But it had disc brakes. And it sold well.*

*Offered initially only as a convertible or coupe, the XK 150 seemed a more mature car, but one still thrilled by its Le Mans lineage* (lower left). *Stolid rivals like the Mercedes-Benz 300SL* (opposite middle) *had Coventry pondering some conservative XK 150-era studies* (opposite, top and bottom).

impressive stoppers at all four wheels, complete with vacuum power assistance.

A good move, that, but late. Disc brakes might well have been bolted onto the XK 140 earlier—as many owners did later—and it was sad that Jaguar lost out to rivals on a little of the prestige benefits from all its hard development work. As early as 1954, the Austin-Healey sports car was available with disc brakes, albeit for the front wheels only, and Triumph followed suit in 1956. That same year, the small British Jensen concern offered them on all four wheels of its model 541, and Citroën in France coupled discs with oleopneumatic power assistance on its front-wheel-drive DS 19, a four-door sedan. But better late than never, and a lot of Jaguar's European and American rivals took years longer to take the disc-brake plunge.

This new-with-the-old mechanical character of the XK 150 was perfectly reflected in its appearance. Many of the basic body lines were the same as the 140's; as were some of the actual dies employed for the more intricate ones. The wider aluminum hood was actually the original 140 piece—albeit sawed apart so a spacer strip could be welded in. The added width appeared in the cockpit, where there were four extra inches of room at the shoulders (up to 50 from 46), and fully six more at the dashboard (to 48 inches).

Hinting at the extra space inside was a higher waistline. If that made for a stodgier look, perhaps the new one-piece curved windshield and much larger backlight were some consolation. Another controversial touch of modernity was elimination of the traditional walnut dashboard trim in favor of leather. Controversial, because

Jaguar was to find that some of its more ardent fans would always demand the wood. But in the XK 150, the company produced a car that looked quite different, yet was easily spotted as a Jaguar—and did it without spending a lot of money.

Initially, however, the 150 was offered only in luxurious coupe or convertible forms, almost as if Jaguar were testing reaction to this obviously fatter cat. When it arrived some nine months later, the starker roadster was bereft of its elbow-cooling cutaway doors, and purists complained. It would prove to be the least popular of the three models, accounting for less than a fifth of the total series production of about 9400. Further, the optional Borg-Warner automatic garnered more orders than before, a sign of the changing sports-car times.

The 150 would receive a number of mechanical running changes during its four-year lifespan. Three of particular note involved the brakes, where a design improvement made replacing pads much easier; the rear springs, where slippery nylon strips sandwiched between some of the leaves reduced their inherent friction-damping; and the rear axle, where a "Powr-Lok" limited-slip differential became available. But the biggest developments involved the engine.

Or rather, engines. To begin with, the familiar 3.4-liter XK six was offered in two states of tune, basic 190 horsepower and 210 bhp, again as on the 140. But the high-performance version now came with a slight but significant change in the cylinder-head department. Replacing the splendid C-type head used on the 140 was a new design called B-head, another example of Jaguar's odd ways with nomenclature. B for "backwards step," maybe.

Or maybe for the blue paint that identified it.

Whatever the reason, the B-head retained the C's larger intake valves, but went back to smaller intake ports in order to speed up gas flow at low rpm. The effect was to fill in the flat spot some critics had noted in the 140MC's power curve. Actually, what Jaguar did was to reshape the torque curve Detroit-style. Instead of 213 pounds/feet peaking at 4000 rpm, the B-head engine offered 216 at a tractor-like 3000. And though peak power was unchanged, it was developed at 5500 rpm instead of 5750. That was according to factory literature. According to people who knew both 140 and 150 intimately, the C-head loved to rev, while the B-head unit didn't like winding much above 5000.

If this slight softening—borrowing a little sparkle from the top to add punch to the bottom—sounds like a sedanish sort of a move, it was. In fact, the B-head came straight from the Mark VIII saloon. But it would be hard to prove that Jaguar lost any sales because of it. Besides, engineers already were preparing yet another head that would fully compensate the hard-core enthusiast.

This new cylinder head was known not by any letters, but as the "straight-port." The work of long-time Jaguar collaborator Harry Weslake, who had been chiefly responsible for the original XK head, it was so-called not because its inlet ports were actually straight, but rather straighter than the original design. That improved gas flow. So did an extra carburetor. Instead of the usual pair of SUs with one-and-three-quarter-inch barrels, the new head wore a new intake manifold to deal with three SUs, each opened to two inches. As an identifier, the straight-port head was painted gold.

Initially a roadster option, the straight-port six was announced in early 1958 as the S-specification 3.4, thus making an XK 150S of cars so equipped. With

*Gold paint identified the 250-bhp "straight-port" 3.4 six. The XK 150 retained the 140's small rear seats, but its instrument surround was now of leather, not wood.*

its better breathing plus 9:1 compression, stronger bearings and clutch, lighter flywheel, and a second fuel pump to handle the increased demand, it was rated at 240 lbs/ft at 4500 rpm and a whopping 250 bhp at 5500—early D-type power in a luxurious road car. It represented a power increase of 56 percent over the 160 in the original XK 120, and, of course, was achieved without increasing displacement.

But a displacement increase did come about soon after, although when this bigger-bore 3.8-liter engine appeared in a production Jaguar later in 1958, it had already seen action in the new Mark IX sedan. The sports car received the 3.8 as an option a year later, meaning model-year 1960, the last full year for XK 150 production. As in the D-type racers of 1957, the capacity increase came via a four-millimeter bore-out, with a revised block casting for better cooling and pressed-in liners (dry, this time) to forestall cracking.

Like the 3.4, the 3.8 was also offered in two states of tune for the 150. The hotter version, again called S, had the straight-port head, triple two-inch carburetors and 9:1 pistons, which made for a rollicking 265 bhp at the same 5500 rpm. Torque went up to 260 at 4000. The normal 3.8 delivered 220 horses at a more sedate 5000 rpm and 240 lbs/ft at 4000.

Jaguar's latest two-seater—really a 2+2 in coupe and drophead form—was even less of a traditional sports car than before, but nobody seems to have minded much. In *Autosport*'s view, the XK 150 went "a stage further towards realizing the modern ideal of the sports car . . . In the not very distant past, noise, discomfort, and a certain amount of temperamental behaviour were regarded as inseparable from the machine of abnormal performance. Jaguar Cars, Ltd., have already proved that these shortcomings are as out of place in a speed model as in a large and luxurious saloon."

*The cabin was wider, but the XK 150 convertible's high waist* (this page) *lacked the previous open car's panache. Overdrive was engaged via a floor lever* (left) *rather than the previous dash toggle.*

That opinion, penned by technical editor John Bolster, went on to extol virtues besides the "immense speed" of the 150: "[I]t is more roomy and practical as an everyday vehicle than its immediate predecessors, with improved all-round visibility . . . The engine has better low and medium speed torque, which increases the already excellent top gear flexibility, but most important of all is the adoption of disc brakes, which really permit the full performance to be exploited."

*The Motor* was likewise impressed with the increased interior room, pointing out the added comfort and convenience of having more width available at shoulders, hips, and feet. *Road & Track* was not alone in praising the improved vision through the wide, curved windshield, which was not so wide and so curved as to hinder one's getting in and out. *The Autocar* made much of the broad-shouldered torque curve, which allowed "a freak demonstration" of a top-gear start. This "turbine-like flexibility," continued *Autocar*, made it a "restful" car to drive on the open road, and also to "trickle smoothly and economically through traffic without the need for constant gear changing . . ."

No, the XK 150—heavier by at least 100 pounds than the XK 140, softer-riding because of the weight, more laden with creature comforts (the roadster even had wind-up windows now), and more practical in every way—was far distanced from the original "Super Sports" concept. And though it had maintained its high value-for-dollar reputation, by 1957 it was definitely falling behind in the suspension, roadholding, and handling departments. Nor were many rapturous paeans of praise being penned about its looks. But none of that seems to have counted against the 150. The car charmed people on its own terms.

And it was still a very fast car. In June 1960,

*Lyons eventually released an XK 150 roadster* (both pages), *which recaptured some of the spareness that hard-core sports-car lovers held so dear. But it was the poorest-selling model, confirming once again Sir William's marketing instincts.*

*Autosport* printed these performance figures derived from Bolster's test of a 3.8S: 0-60 in 7.6, quarter-mile in 16.2 at 91 (extrapolated from the graph), and a top speed of 136.3 mph. Acceleration had been handicapped by an apparently slipping limited-slip, while the top-speed runs were done with foglamps and radio antenna in place and on the normal street tires. Given some special preparation, the car was likely to be capable of 140.

"This is an enormous velocity for a luxurious touring car," observed the experienced Bolster. "There is a wonderful sensation of having almost unlimited power under one's toe, and provided that this immense reserve is never abused, it can be regarded as a very real safety factor in time of trouble."

His most telling point: "A Jaguar driver ought to be a good driver, because he need never be in a hurry."

On this note ended a long line of model development that actually began more than 20 years earlier with William Heynes' prewar investigations of independent front suspension systems. The original XK 120 of 1948, conceived as one kind of sporting machine, had "grown up" to assume a very different persona. By the beginning of 1961, when XK number 30,357 was driven off the Browns Lane production line, the 150 still was a good car, still an effective and well-liked car.

But perhaps as important, the 150 was the biggest sports-car credit ever on its maker's balance sheet. Though its 40-50 weekly production rate was piddling by Detroit standards, it was enough to help Coventry keep taking care of business. Moreover, the roadster's relatively poor sales showing suggested that Lyons had correctly read the changing market. With the amazing new sports car it was about to spring, Jaguar would reap even greater fame and fortune.

*A 3.8-liter six was made optional for '60, the XK 150's last year. In S tune it had 265 bhp, good for 0-60 mph in 7.6 seconds, a top speed of near 140, and the "wonderful sensation of having almost unlimited power under one's toe."*

CHAPTER XII

# E-TYPE: IGNITING A NEW SPORTS-CAR REVOLUTION

To anyone for whom the automobile is more about romance than utility, the Jaguar E-type ranks among the most important cars ever created. And not only for its virtues as a vehicle. Yes, in itself it was a very exciting sports car, combining speed with style, savagery with civility. But then, Jaguar devotees had come to expect that from their marque.

The E-type was something more—much more. There are enthusiasts who hold that racing is the highest of the automotive arts, and thus a roadgoing automobile must be derived from racing experience. For them, a sports car based on a competition car is the best car. Such disciples perceive a natural link between the demands of the speedway and the pleasures of the open highway, and believe fervently that "racing improves the breed." The E-type validated their theology.

Here was a sports car that was not only exquisitely pleasing to look at and exciting to drive in every way, but one sired directly by a racing car. And not just any racing car, but the Jaguar D-type, which had won the world's most prestigious sports-car race three years running, in 1955, '56 and '57. Though the E-type did not appear until 1961, it was quite clearly the D's lineal descendant, an honest and genuine attempt to adapt the Le Mans car's performance technology to everyday use—to tame the racer for the road.

Jaguar did literally that as a first step. Late in 1956, the company began converting the actual customer-version D-type racer into a highway-capable sports car. Labeled XK-SS, it was convincing proof that there was more to the job of taming wild beasts than draping them in harness.

The XK-SS was not exactly a failure as a sports car, but it was one of those unfortunate ones whose failings seem to outnumber their finer points. A "yes, but," sort of car.

To begin with, the project was a ploy, and everybody knew it. By the time the XK-SS appeared, the D-type had completed its third season and the Jaguar factory had withdrawn from racing. It's hard to see it from the vantage point of today, when all such cars are so immensely valuable, but at the time, the D-type was an aging athlete. It still had some good runs left in it—including another Le Mans victory—and was more widely honored than ever, but retirement was looming. Faster racers were springing up all over.

In a frank attempt to wring some added value from an obsolescent design, Jaguar hit on the idea of fitting some "production" D-types with full road equipment. That would suit eligibility rules for the C-Production racing class of the Sports Car Club of America, but meant building at least 100 examples. Jaguar might have a hard time placing that many with true racers. Still, perhaps some customers in the U.S. and elsewhere might be attracted by the prospect of owning a true super-sports street car. Worth trying, anyway.

It should be noted that such a product had been on the collective company mind for about as long as the D-type itself had been in existence. Sometime during 1954, Jaguar's resident aerodynamicist, Malcolm Sayer, had sculpted a lump of modeling clay into a miniature roadster that strongly hinted of both the XK-SS and E-type to come.

The job of making racing D-type into roadgoing XK-SS was done as thoroughly as possible within the constraints inevitable when starting with a purpose-built machine. The center of the monocoque underwent some surgery, having its shoulder width opened up and its cen-

*Its lines libidinal, its performance predatory, the E-type leaped onto the world's sports-car stage at Geneva in March 1961. Like the XK 120 of 13 years earlier, it was a Jaguar masterwork, but one with real racing roots.*

tral brace between the seats removed, as well as a passenger door cut into the left side. Also shorn was the driver's headrest and its fairing, leaving the previously concealed quick-action fuel cap exposed. The low plastic windscreen was replaced with a tall, rather upright one made of glass in a steel frame, and wipers were added. So were a sketchy folding top with detachable side-screens and flimsy quarter-bumpers attached directly to the fragile fenders. The only place to hang a muffler turned out to be along the left side, under the rocker panel. The only place to accommodate luggage was a rack plunked atop the tail. Installing some cockpit upholstery, turn signals, and trim rings around the plastic headlight covers completed the transformation of wild racing beast into a demure, if perhaps overdressed, street beauty.

Nevertheless, the XK-SS retained the full racing powertrain as supplied on customer D-types. This meant a 3.4-liter six with dry-sump oiling system—the regular wet-sump engine wouldn't fit under the low "bonnet" anyway—and a trio of Weber carburetors. Output was unchanged at 242 pounds/feet of torque at 4000-4500 rpm and 250 horsepower at 5750. Performance thus stood to be shattering, as the dual-purpose car weighed only about 100 pounds more than the pure racer. To help move the necessary 100 copies, U.S. price was set artificially low: just under $7000—nearly $3000 less than a pure D-type. Possibly the bargain of the decade.

Announced on January 21, 1957, the XK-SS attracted the two types of customers Jaguar had hoped it would. There were indeed people for whom it represented automotive nirvana. One was Hollywood action man Steve McQueen, who is known to have been delighted with his. But several others were purchased by less famous folk who used them for both racing and vivid

road driving.

How vivid? The XK-SS seldom found its way into professional road-test hands, but *Road & Track* clocked one at 5.2 seconds 0-60 and through the standing quarter-mile in 14.1 at a little over 100 mph. That was despite considerable wheelspin due to the lack of a limited-slip differential, but another example tested by *The Autocar* pretty well confirmed those numbers. The traditional 3.54:1 rear-end gearing of other Jaguar street cars was rather short for the XK-SS, so top speed was limited by the 5800-rpm redline to 144 mph.

In all, stupendous performance for the day—for just about any day, in fact. But enjoying it took some commitment. As the cockpit was basically identical to the D-type's, accommodation was none too generous. The passenger side was especially cramped in foot room. In America, that passenger sat watching the oncoming traffic, because no XK-SS was built as a "left-hooker." The passenger also had to put up with heat beating through the aluminum cockpit side from the left-mounted exhaust system.

The driver had more fun, though it was fun of a demanding sort. While the engine retained much of the tractability for which the XK powerplant was long renowned, it did have racing cams, so its power was concentrated in the upper third of the rev band. Also, because the D-type engine had no specific flywheel, it responded vividly to the throttle. Both characteristics made smooth engagement of the abrupt multi-plate racing clutch all the harder.

There were other functional quirks that rendered the XK-SS a questionable proposition as a true dual-purpose vehicle. Some drivers familiar with the D-type felt that the SS chassis flexed a little by comparison, because it

*Coventry wrung some added value from the aging D-type with the XK-SS of '57. Based on the Le Mans-winning competition car, this was a street machine also intended for production-class racing. At $7000, it was a bargain, but only 16 of a planned 100 were built before fire destroyed its assembly area.*

*XK-SS used the full racing powertrain of the customer D-types. The 250-bhp 3.4 six with its three Webers had both beauty and brawn.* Road & Track *saw 5.2 seconds 0-60 and 14.1 at just over 100 mph in the quarter-mile. Top speed was a redline-limited 144 mph.*

didn't have the cockpit center brace. Then there was that bulky exhaust system, which apparently was pretty noisy and probably not hard to scrape on a curb. Prolonged urban slogging risked running the battery down, because the generator (no alternators yet) was set up to need more than 2000 engine rpm to charge. When the time came to top up the huge rubber-bag gas tank, the same 44-gallon (U.S.) item that filled the shapely tail of the D-type, it had better not be raining, because the top had to be dismantled to get at the racing-type fuel port.

On the good side, ride quality was described as surprisingly comfortable. The windshield seemed to work well, though it was something less than graceful to look at. And though the steering suffered some of the kickback over bumps characteristic to rack-and-pinion mechanisms, it was quite quick at 2.3 turns lock-to-lock on a 32-foot turning circle, yet pleasingly light. There was a noticeable tendency to understeer that had been deliberately built into the D-type for stability at Le Mans, and which proved to work well on the highway. Yet whenever a corner seemed too tight, there was plenty of poke to get the tail out. The all-wheel disc brakes, of course, were fabulous.

Sixteen cars were built up from scratch—more accurately, from incomplete D-type chassis. Two more finished D-types were converted to XK-SS specification, one apparently the very car that had finished second at Le Mans 1954 and was later tested for *Autosport* by John Bolster. (On the other hand, several owners of cars originally built as XK-SSs subsequently converted them to D-type spec.)

Then, overnight, production ended. Was such a super-sports concept too virile to be as popular as one might have expected? Perhaps, but we will never really

*Weighing only 100 pounds more than the pure D-type racer, the XK-SS's performance was shattering, but racing cams concentrated power in the upper rev ranges, the chassis suffered a little flex, and there was no trunk. However, strong disc brakes, a compliant ride, quick steering, and balanced handling compensated greatly.*

know. On the night of February 12, 1957, only three weeks after the XK-SS had been formally introduced, the part of the Browns Lane factory where it was built caught fire. Several incomplete chassis and many parts were destroyed; more crippling, so were most of the jigs and tooling.

Thanks to enormous effort and dedication, production of normal cars was going again within days, but it just wasn't worth restarting the XK-SS line. So in a flash, Jaguar was out of the super-sports business, though perhaps that wasn't entirely bad. But there was a lot of obvious good in the basic idea of a sports car built on D-type lines, and it was worth pursuing.

Thus, step two toward the eventual E-type. It was taken late that same year, 1957, when Jaguar's engineering department laid down an experimental chassis for a genuine roadgoing sports car, a project designated E1A.

Whether the "E" simply stood for "experimental" or had already been chosen for a new production model is not clear. The meaning of the "1" is clear, of course. The "A" signified "aluminum," or perhaps "alloy," the material of which both body and chassis were made. That implies Jaguar was already envisioning a later version of steel, which was more suitable for volume production.

Outwardly, the E1A prototype bore a strong resemblance both to the D-type/XK-SS and to the clay-model roadster that Sayer had carved back in 1954. Structurally, it was very much a D-type derivation, with a similar ovoid-section central monocoque tub and space-frame front structure. As on the earliest Ds, tub and tubing were welded together, with the front frame again being light alloy. But there were two departures from D-type convention. Instead of a 3.4/3.8 engine, E1A carried the short-stroke XK six of Jaguar's then-new 2.4-liter compact sedan. This was because the prototype was physically small, being shorter, narrower, lower, and probably also lighter than the race cars that came before and the road model that was to come. Also, E1A had an experimental independent rear suspension instead of the D's live rear axle.

It's worth remembering that Jaguar engineers had been working with independent suspension since before World War II, beginning with William Heynes' early investigations. During the war years, the company had built two different prototypes for a lightweight military vehicle, both with independent suspension of all four wheels. The XK 120 and Mark V sedan emerged in 1948 as the first production Jaguars with separately sprung front wheels. Later, during the D-type program, some testing had been done with a de Dion rear end, which has some of the advantages of full independence. So the thought had long been in mind to bring Jaguar's road cars, both sports models and sedans, into the modern all-independent world. Late 1957 was the time, and E1A was the development vehicle.

It was running by early 1958. And run it did, logging many thousands of hard miles on test tracks, race tracks, even public highways. In fact, on one extraordinary occasion, Sir William Lyons handed the keys of the top-secret prototype over to a member of the automotive press—the editor of *Motor*, no less—who was to take the little light-green roadster along some favored back-country roads in Wales and report back. He returned with words like "astonishing," "sensational," and "world-beater."

The scribe kept faith with Lyons and kept quiet about the car in public, but sent a "secret and confidential" memorandum to his boss in May 1958. Published many years later by Paul Skilleter in *Jaguar Sports Cars*, the note revealed this editor's understanding that the production sports car to come would be a 3.0-liter with an amazing output of 286 horsepower and a projected top speed "not very far short of 150 mph, which is going to make us think." Jaguar was thinking of making 100 per week, he added, and said that the new model, which people around the factory were already referring to as "the E-type," was to go on sale in the autumn of 1958.

Alas, that timetable would eventually be put back two-and-a-half years. Happily, most of the rest of his predictions came true.

Around the time Jaguar was playing with the E1A, it also built the E2A. Though similar to the E1A in both looks and rear suspension concept, it resembled even more the eventual E-type. It even had the longer, 96-inch wheelbase of the forthcoming road car, although in the interest of high straightaway speed it had a narrower

*A bridge between the XK-SS and the production E-type was the E2A, shown here being readied for test runs at Le Mans in 1960. Made of aluminum and powered by an XK six of nearly 3.0 liters, it set best practice time, but couldn't finish the race.*

track than the D-type. Actually, E2A was conceived more as a racer than road car. Many in the company still hoped Jaguar would return to formal competition, and this other "E-type" was really a follow-on to the D-type.

And after some years of clandestine development, the E2A finally did get to race, being taken to Le Mans in the spring of 1960 as an official entry. In fact, it was an official entry of the Briggs Cunningham team along with a trio of Chevrolet Corvettes (one of which would finish eighth overall and first in the Grand Touring class). Word was given out that this spectacular new Jaguar had been "specially commissioned" by the marque's American stalwart, but the truth was the aging hack had already been pushed to one side when Cunningham was asked to take it racing. For a pre-race test session at Le Mans in April, it ran sans paint, its bare aluminum bodywork bearing only a British registration number. By race week, though, it was in American racing livery of white with blue stripes. But nobody could have doubted its real origin and original intent.

Made mostly of aluminum like the D-type, the E2A "E-type" ran at Le Mans with a special aluminum-block XK engine featuring dry-sump oiling and fuel injection. Nearly square bore and stroke dimensions of 85×88 mm meant displacement of 2997 cc, just 3 cc below the contemporary 3.0-liter limit in the prototype class where the E2A qualified. With big valves, hot cams and a high 10.0:1 compression ratio, horsepower was a tingling 293 at a whizzing 6750 rpm. Indicating the peaky nature of this engine's power delivery was the high 6000-rpm speed for peak torque, which was 230 pounds/feet.

This "E-type" proved quite fast in test sessions and in the race itself. Driven by American aces Walt Hansgen and Dan Gurney, it even set the best practice time. At the end of the first 8.36-mile lap, the E2A came by the pits in third place behind a Maserati and a Ferrari, and well ahead of the other Jaguar entered, a D-type, that grand old model having its last Le Mans hurrah. Then a broken fuel line set up a series of long pit stops to correct engine troubles. E2A made it past the ninth hour, but eventually quit. *Autosport* thought the Jaguar had been strong enough to win outright, and expressed regret that there hadn't been a team of three.

Cunningham then took the car back to the States, where he raced it several times with a normal 3.8-liter, Weber-carbureted engine crammed under a "power bulge" in the hood. In this form, Hansgen won with it at Bridgehampton, Long Island, beating a Jaguar-powered Lister.

But the mid-engine revolution was at hand. At least the E2A would find safe haven in the hands of a collector. The E1A? Unsentimentally, Jaguar cut it up as

*The E2A* (left) *competed in Europe and America, but the front-engine roadster had no place among the new breed of mid-engine racers. Rather, its main job was to help Jaguar develop the E-type* (below). *The family resemblance is obvious.*

scrap—though perhaps only because the company had long since turned its full attention to preparing a roadgoing E-type for production. Apparently, the basic concept had evolved since *Motor*'s clandestine evaluation of the E1A in Wales. No longer a very small, very light sports car, it had grown to the wheelbase of the E2A (96 inches) and track dimensions at both ends as wide as the D-type's front (50 inches). Chassis and body would be made of steel rather than light alloys, and though a lift-up hood/front-fenders assembly would be retained, it would reveal a normal iron-block 3.8 with a standard deep sump.

Basic chassis structure again involved a front space-frame of square-section steel tubes, but this was now bolted to a rear monocoque of sheet steel. Front suspension would be very similar to the D-type's, with forged control arms, tubular shock absorbers, and longitudinal torsion bars. Steering would be rack-and-pinion, of course.

Such familiarity only highlighted what would be a big attraction of the forthcoming E-type: Jaguar's latest independent rear suspension. This employed the usual U-jointed halfshafts to take power from a chassis-mounted differential out to the wheels, but also used them as upper control links. This efficient, two-jobs-in-one idea was not new, having already been used by Lotus on some of its racing cars as well as on its roadgoing Elite. In America, Zora Arkus-Duntov had been using such a design experimentally since the late 1950s and would adapt it for the 1963 Corvette Sting Ray.

Under the halfshaft was another control arm forming the lower element of a wheel-locating parallelogram. In the E2A, this had been a wide-based transverse "wishbone" or A-arm, but in the production E-type it took the form of a lateral tube with a forged yoke on each end. Its job was to locate the wheel in both camber and toe. A third control link ran from this tube forward to the bottom of the chassis to resist braking and accelerating forces.

"Ears" on each tube supported a pair of vertically sited coil-spring/tubular-shock units per side. Having pairs precluded twisting forces on the tubes and allowed the spring/shock units to be small enough to fit beneath the rear floor area without undue space intrusion. That was important, because this same independent rear suspension was also going to be used in Jaguar sedans. The entire suspension package was carried on a steel substructure, which itself was rubber-mounted to the main mono-

*Unlike its XK 140/150 predecessors, the production E-type came into the world as a strict two-seater only in convertible or coupe form. Clearly evolved from Malcolm Sayer's less-harmonious E2A, the E-type's elongated purity of line owed much to the inspired hand of born-stylist William Lyons.*

coque to keep noise and road shock out of the cabin.

Naturally, all four brakes would be discs, but the rears would be mounted inboard, next to the differential. Some of the time between 1958 prototype and 1961 production car had been spent making sure that heat from the rear brakes wouldn't soak into the differential.

By contrast, Jaguar decided to simplify things in the engine department by fitting just the 3781-cc six. Moreover, it would be offered only with the straight-port head and its trio of two-inch SU carburetors, as on the hottest XK 150S. Although 8.0:1 compression was specified for markets with poor-quality gas, most E-types would run 9.0:1 pistons, in which case the 3.8 was said to produce 265 bhp at 5500 rpm and 260 pounds/feet of torque at 4000. Unlike its canted position in the D-type, the engine stood up straight here.

A four-speed manual would be the only available transmission. No overdrive, and no automatic. This wasn't for the sake of simplicity, though: There was simply insufficient room for those options in this compact, tightly packaged sports car. However, a limited-slip differential would be standard.

With all its civilizing changes, the old XK roadster had become a contradiction in terms as a 150, and was not selling well anyway, so E-type body styles would be limited to the familiar pair of coupe and convertible. The latter was occasionally called Open Two Seater, but in neither model was any attempt made to cram extra seating in the back, "+2" or otherwise. For the first time, Jaguar offered the drophead with a detachable hard top, an idea doubtless borrowed from America and here also rendered in fiberglass.

Perhaps not surprisingly, the new production sports car emerged looking much like the E2A. But where the latter had seemed a bit stubby, perhaps even dowdy to eyes long used to feasting on the D-type, the roadgoing E was so lovely as to make the heart ache. The original lines may have come from Malcolm Sayer's wind tunnel, but in the grace, the balanced proportions, the subtle electric excitement, one surely saw the hand of William Lyons.

For its formal introduction to the world, the E-type was taken to the international arena of Switzerland and the Geneva auto show of March 1961. It was fully the sensation the XK 120 had been in London over a dozen years before. But even more so. Incredibly, Jaguar had ignited yet another sports-car revolution.

*Dashboard toggle switches and prow unsullied by heavy bumpers bespoke a time before stringent safety and crash standards. But they also were vital details on an automobile that was and always will be more about romance than utility.*

CHAPTER XIII

# E-TYPE: ON THE ROAD

If you were the kind of driver for whom the E-type Jaguar carried a message, you approached the car with a sense of awe bordering on reverence. For here seemed the fulfillment of a long-held dream: a roadgoing racing car.

Packaged in one wind-piercing projectile was a race-proven twin-cam engine with three carbs, a four-speed gearbox, a limited-slip differential, fully independent suspension, all-wheel disc brakes (still inboard at the rear!), rack-and-pinion steering, and a lightweight chassis

*Where sex appeal came for lessons. "The greatest crumpet collector known to man" crossed the line from provocative to suggestive. And did it quickly.*

featuring a fusion of space-frame and monocoque technologies. The E-type was the child of the D-type, three times a winner at Le Mans, and looked it. Better, the E-type performed almost like the D-type.

Even so, its great prowess came with all the silky smooth sophistication customary in Jaguars, that feline merging of muscle with manner that left no one unmoved on levels both intellectual and intestinal. Yet even for confirmed Jaguar enthusiasts, this car was something new.

Twelve years of familiarity with the XK-series had bred a feeling in some quarters that Jaguar's two-seaters were physically too large and heavy to be "true" sports cars. Yet against the 150, especially, the E-type seems almost petite. Despite that long nose, it is an inch-and-a-half shorter overall, and six inches briefer in wheelbase. Its rounded bodywork is an inch wider, but track measures an inch-and-a-half narrower. Matched roofline to roofline, the E-type coupe stands a remarkable seven inches lower. And it weighs several hundred pounds less than the closed 150.

This impression of compact efficiency continues as you open the very short door and fold yourself into its narrow opening (fewer than 21 inches separate windshield post from latch pillar). The sill is relatively high off the ground, and a bulky seven inches wide. At a diameter of 16 inches, the steering wheel commandeers some of the space your legs want as they slide in.

You are comfortable enough once settled, but the D-type's "performance-first" racing philosophy is very evident in the snug E-type cockpit. Within the shorter wheelbase, the engine sits farther back than in the XK 140/150, and while interior volume is adequate, it is hardly generous. In fact, shoulder and head room are each an inch tighter than in the 150 coupe. However, the instrument panel and steering wheel seem farther away. The wheel still adjusts for reach via a friction collar on the column and, reviving a good idea from the classic SS 100, can be adjusted for rake via a wrench. Both column and pedals angle away from the engine, so you are immediately conscious of sitting with your legs slightly askew and with the wheel canted in your hands.

The wheel itself appears high-set, but is lovely to behold. In the style of the day it is large across but thin in the rim, made of polished wood and aluminum, and has racy "lightening holes" drilled into its spokes. Those spokes are quite springy, though, which can allow a disconcerting amount of flex in your hands. The seat seems minimal: thin in the cushioning and short in the backrest. Because you no longer perch atop a separate chassis, as in the XKs, you feel nestled much lower relative to both macadam and machine. The gearshift knob now stands almost as tall as the flat horn boss in the center of the steering wheel, and is only a handspan away from the rim.

For the first time on a Jaguar sports car since the SS 100, primary instruments—rev counter and speedometer—are located straight in front of the driver. And the tach has been changed so the needle now rotates in the same clockwise direction as the speedo.

Peer through the wide but shallow windshield—so shallow it requires three short wiper blades for coverage—and you are captivated by the sensuous curves of the hood. There is enough bodywork falling away out of sight to make you worry a little in tight quarters, but then, this car was not born for tight quarters. That beautiful bulge rising so prominently in the center of the "bonnet" tells the tale; it seems to be pointing the way to the far end of the unimaginably fast Mulsanne Straight. Or to any horizon you care to imagine.

A push of the starter button and the cat awakes with a spine-tingling growl, that classic Jaguar sound. At once suave and savage, it conjures the image of Tarzan in

*Jaguar returned to truer sports-car dimensions with the E-type. It was shorter than the XK 150 and seven inches lower. The coupe was several hundred pounds lighter than its closed predecessor and had a rear luggage door that opened to the side. The 1961 U.S. list price was $5595.*

*The nose section canted forward revealing the 3.8 twin-cam six. It made 265 bhp at 5500 rpm and mounted further to the rear than in the XK 150. A four-speed manual was mandatory. The cabin was inviting, but not spacious.*

a tuxedo. With effortless ease, the long-stroke engine seems to pick the car up and move it down the road in an oily-smooth feline lope. As your foot goes down harder, the growl changes to a predatory snarl. It is a true warning. Although there is plenty of muscle at lower revs, at about 3500 rpm the 3.8 engine takes a deep breath, the exhaust note becomes a wild howl, and the beast seems to leap for the redline. No red-blooded enthusiast can possibly have a soul so dead as to not feel exhilarated by the XK six.

Even magazine writers had trouble maintaining professional objectivity. One of the first to fail was John Bolster, whose factory-press-fleet E, a Fixed Head Coupe, was delivered with the promising British registration plate 9600 HP. His report ran in the issue of *Autosport* appearing in England on Friday, March 17, 1961, two days after the E-type was first revealed and just as it was making its debut in Switzerland. "If *Les Vingt-Quatre Heures du Mans* has been responsible for the new 'E' Type Jaguar," wrote Bolster, "then that Homeric contest on the Sarthe circuit has been abundantly justified." Bolster had always written favorably of Jaguars, but he strained for superlatives in writing of this one.

"Here we have one of the quietest and most flexible cars on the market," the writer rhapsodized, "capable of whispering along in top gear at 10 m.p.h. or leaping into its 150 m.p.h. stride on the brief depression of a pedal. A practical touring car, this, with its wide doors and capacious luggage space, yet it has a sheer beauty of line which easily beats the Italians at their own particular game . . . All this, and 20 m.p.g. [17 U.S.] economy too, comes at a price which is about half that of the current crop of glamour-wagons. It is all the more remarkable that this Grand Touring car *par excellence* is directly descended from the Le Mans-winning 'D' Type. The Jaguar engine is one of the most amazing achievements of modern times. . . ." Some people are hard to please.

But "wide doors"? Well, Bolster would probably have crawled in through a keyhole to drive this car. "To give some idea of the potential performance," he explained, "let us try to imagine an even lighter XK 150S with a greatly improved aerodynamic penetration. Let us

*Three short wiper blades were required to cover the shallow windshield. The shape of the nose intake was perfection itself, but glass-covered headlamps provided inadequate illumination. Body rigidity was high for a convertible.*

then envisage the virtual elimination of wheelspin by virtue of the I.R.S. The result, of course, is something out of this world. . . ."

By "I.R.S.," Bolster meant the E-type's new independent rear suspension, which he counted a tremendous value. One performance limitation of both the D-type racer and the hotter-engine XK sports cars had been their "live" beam-type rear axle. "Live" referred to the method by which it was mounted, but drivers of higher-performance Jaguars would have been forgiven for thinking it was called "live" for the way it bounded about on bumps or tramped violently when too much power set a wheel spinning. All that was history in the E-Type. Observed Bolster: "*Autosport* has always championed the independent suspension of the rear wheels, and we have often wondered what a Jaguar would be like with this desirable design feature. Now we know, and the answer is—it's a winner!"

Acceleration, Bolster enthused, was "almost incredible . . . In the past, only racing and the larger sports/racing cars have been capable of breaking 15 seconds for the standing quarter-mile. The 'E' Type actually clocked 15.1 secs. on a wet road and 14.8 on an almost dry one. . . ." His 0-60 time was 6.8. For top speed he recorded 148.1 mph, although "a strong side wind was blowing at the time, and I am happy to call this a genuine 150 m.p.h. car." A velocity of 145, he continued, "came up from time to time during ordinary road motoring." He did explain that to see these speeds with the standard 3.31 axle ratio, one had to venture beyond the 5500-rpm redline, which Jaguar had told him was "permissible" in top gear. "I found that 5,800 r.p.m. was very quickly gained, the last 200 r.p.m. taking rather longer to come up."

Powerful disc brakes and steady, responsive handling made such speeds seem reasonable. "The wet road behaviour is indeed excellent, a slight initial understeer being quickly converted to a balanced condition or a tail slide, depending on the position of the accelerator pedal. On dry roads, the most outstanding feature is the machine's behaviour on fast, open bends. Quite appreciable curves may be rounded at 110 m.p.h., and full throttle may be held without any tendency to 'lose it.'"

Bolster also made the significant point that Jaguar's painstaking development of a rear subframe mounted on rubber bushings had eliminated one of the chief arguments against independent rear suspension: noise intrusion into the cockpit. The E-type, he said, was quieter in terms of gear whine and road rumble than most conventionally suspended cars.

For his part, *Autosport* editor Gregor Grant had this congratulatory capsule for Coventry: "It is a true conception of a modern sports car, displaying features which could only have resulted from intensive development work and a racing history. . . ."

Similar sentiments filled public prints all over the world. Seldom has any new car received the voluminous coverage lavished on the E-type during 1961. Not only was it sexy and sensational, ranking right up there with recognized supercars, it was attainable. At $5595 in the U.S., it was less than half the cost of a contemporary Ferrari. As with the XK 120, many thousands of people would be able to buy the E-type. And that made it important.

Five days after British readers had devoured *Autosport*'s test, *The Motor*'s came out. It told a tale of 3000 miles in England and on the Continent with an Open Two Seater. The staff had anticipated that the E-type would be "something of a landmark in sports-car progress" but they were resoundingly convinced after their week's hard motoring: "Curiously enough, its very close connection in design and appearance with competition Jaguars gives people the impression that this is essentially a racing car with all the limitations for ordinary use that this implies. Nothing could be farther from the truth; admittedly, it is quite easily the fastest car ever tested by *The Motor*, but the roadholding is entirely capable of handling the power, the springing is more comfortable than that of many sober touring cars and the engine is extremely flexible and devoid of temperament. The ease and delicacy of control is such that 220 b.h.p./ton was no embarrassment at all on the packed snow and ice of Swiss mountain passes using ordinary racing tyres."

"Ordinary racing tyres"? Well, the manufacturer did list Dunlop R5s as an optional extra. With these better

*Gauges straight ahead, gear shift a handspan from the quintessentially sporting steering wheel, bonnet bulge aimed down the Mulsanne: Jaguar!*

boots, plus a weight advantage of about 100 pounds over the coupe, *Motor*'s roadster motored to 60 in a breezy 7.1 seconds and through the standing quarter in 15.0 (a retest on normal Dunlop Road Speeds dropped that to 14.7). Maximum speed was 149.1 mph with the soft top up. Here, too, the engine was taken 500 rpm into the red zone. The cost of all this speed was a "touring fuel consumption" of just under 18 miles per U.S. gallon.

As a driving instrument, the E-type was found to be a delight in many ways. The independent rear suspension was able to give a soft ride over rough roads allied with a feeling of great stability. The steering was light, smooth and precise "to an outstanding degree," yet there was little kickback. Body rigidity was deemed high for a convertible.

As for handling, *Motor*'s description likely set a million mouths to watering: "A great deal of clever development must have been required to produce cornering characteristics which are not only outstandingly good but particularly well suited to the unusual power-to-weight ratio. It is basically very near to being a neutral steering car, but the driver is constantly astonished by the amount of power he can pile on in a corner without starting to bring the tail round; as with front-wheel drive, hard acceleration through a bend is the right technique, and lifting off suddenly gives a marked oversteering change. Naturally, the power technique can be overdone in the lower gears, but this merely increases the nose-in drift angle in a most controllable way. It is possible to go on increasing the sideways 'g' value to a quite surprising level, because the E-type retains its balance far beyond the point at which most sports cars have lost one end. The very low build (we only realized how low when we saw a small foreign GT coupe towering over it) and anti-roll bars at both ends keep the roll angles right down, and it seems natural to throw the car about in a manner usually reserved for smaller and lighter sports cars."

*Motor* noted some negatives, though. There was some audible engine "pinking" (ping, or detonation) in the 2000-2500 rpm range even with 100-octane petrol. Oil consumption was very high, about one American quart for every 250-300 miles. Spark plugs tended to protest

*Putting the spur to one of the very first E-types, writer John Bolster saw 0-60 mph in 6.8 seconds, the quarter-mile in 14.8 on a damp road, and a top end of 148. Alas, testing also revealed high oil consumption, spark-plug fouling, and a balky, heavy shift linkage made worse by long, slow clutch travel.*

prolonged low-speed work by fouling. Then there was the four-speed transmission: slow, heavy, balky, and noisy-shifting, exacerbated by Jaguar's typical long clutch travel, coupled here with incomplete clutch disengagement. Pedals weren't well spaced for heel-and-toe techniques, either.

The magazine also opined that by sports-car standards, the coupe offered ample luggage space (enhanced by a side-hinged hatch door), but judged the roadster's "boot" too shallow to be very useful. Wind noise was judged excessive at the junctions of the side windows and top, while seats were termed "unsatisfactory" because of inadequate lumbar support and a tendency to let the body slide forward (this well before seatbelts became common). Taller people found the cockpit too short and too low. Ventilation was not up to hot weather, and fuel odors sometimes invaded the cockpit. The editors said the aerodynamic "Le Mans-type" glass-covered headlights proved "not really adequate for the performance. There was insufficient spread to illuminate the sides of twisty roads and the dipped beams seemed to cause considerable annoyance to other road users."

But overall, thought *The Motor*, none of the bad points were likely to overwhelm the good ones in the eyes of Jaguar's clientele: "It is difficult to see . . . how this car can fail to be a tremendous success. The sheer elegance of line which Jaguar seem able to produce by total disregard for fashion trends is allied to a combination of performance, handling and refinement that has never been equalled at the price and, we would think, very seldom surpassed at any price."

*Motor Sport*'s William Boddy had a much shorter run of only two hours in an E-type coupe (apparently Bolster's 9600 HP), but it was long enough to make him a fan: "Put the E-type in top gear and it just goes faster and faster, until it is cruising along M1 [Britain's first superhighway] at 6100 rpm, which we calculated to be a pretty genuine 155 m.p.h. At this speed it is possible for driver to converse with passenger in normal tones, wind noise being low and little noise coming from transmission or final-drive—a fantastic experience!" Any speed below about 110, Boddy said, "is loitering." Among superlatives in the balance of the story were "exceptionally" and "outstandingly"; "staggering" he used twice.

This same well-flogged coupe carried famed competition driver and longtime Jaguar friend Tommy Wis-

*"The Jaguar engine is one of the most amazing achievements of modern times . . . [in] one of the most flexible cars on the market, capable of whispering along in top gear at 10 m.p.h. or leaping into its 150 m.p.h. stride on the brief depression of a pedal..."*

dom, plus wife "Bill," down to the Alps in 1962. He returned with some interestingly philosophical remarks for readers of *The Motor*. The E-type's sheer excellence, Wisdom felt, pointed up a very real, very fundamental change that had occurred in sports cars since the 1930s—sports cars such as the SS Jaguar 100. As he wrote after reaching a speed of 120 mph across France, with plenty more in reserve, "I could not rid myself of the recurring thought—the racing Jaguar 3½-litre S.S. 100 at Brooklands before the war was not as fast as this."

Yet even at such speeds, Wisdom went on, the E-type demanded a conscious effort lest one lose concentration: "Many of these new cars, because they are so quiet and comfortable, and hold the road so well, need a new approach to driving. Though these modern cars are intrinsically safer, they can, in inexpert, inexperienced hands, be more dangerous than the old-type sports cars. Think back to the . . . S.S. 100 and other great machines of the 'thirties. Their noise, both exhaust and mechanical, the harsh suspension, uncertain brakes, heavy, direct steering, the very rush of wind at a mere 80 miles an hour made you concentrate—mentally and physically—on handling them. The rattling, jarring, bouncing machine really kept you on the job."

With this perspective, he felt that the "faster, quieter, safer machines of today can easily lull you into a false sense of security. Unless you 'Drive on instruments' you may be chatting away to your passenger with too little appreciation of the speed at which you are proceeding. It is too easy. Stirling Moss once said that concentration is the hardest lesson to absorb in the whole manual of driving instruction, and the very effortlessness of the modern fast car makes this more difficult."

Jaguar, of course, had been among the most aggressive and successful automakers in bringing this about. It was hardly fair to blame the company now for doing too good a job. Realizing this, Wisdom concluded his essay by saying of the E-type, "Here is a car which, like a pedigree gun or a green-heart trout rod, is so worth learning to use properly."

These words joined millions of others. Everyone, it seemed, had something to say or write about the "staggering" E-type. Thus was fulfilled the prophecy of that *Autocar* editor, uttered after his clandestine test of the experimental E1A: Jaguar's new 150-mph super sports car really *did* "make us think."

*Lessons learned at Le Mans helped imbue the E-type with balanced handling and communicative steering, its new independent rear suspension contributed to fine overall control, and disc brakes—inboard at the rear—inspired confidence. Performance that rivaled the Italian exotics and styling that to some eyes beat them—at half the price. The E-type was genuine Lyons value.*

CHAPTER XIV

# E-TYPE SERIES 2: GRACEFUL AGING

As did the XK 120 before it, the E-type validated its performance claims by winning the very first race it ever contested. And it did so in a thoroughly convincing way. On the sunny spring afternoon of April 15, 1961, after 25 quick laps of the tricky little Oulton Park road course in Cheshire, future world champion Graham Hill took a race-prepared roadster to victory over a field that included an Aston Martin DB4GT, another new E-type, and a pair of Ferrari 250GT short-wheelbase berlinettas.

On such a resounding note began a long and impressive competition career for Jaguar's new sports car. Impressive, especially, in that nothing about the showroom E-type was ever meant for racing. It was primarily a road car that made good use of design principles proven in racing—such good use that it could go to the track and, often, beat far more exotic and expensive cars.

Beauty, exciting mechanical specification, rave reviews, and now a growing record of competition success; what possible excuse not to buy one?

Well, even those most captivated by the cats from Coventry have been willing to admit over the years that no Jaguar was ever without flaw, especially new ones in their infancy. Yet as the first E-types went out to prowl on the roads of the world, Jaguar was drawing up a job-list of things that needed improving based on customer feedback and the factory's own experience. Some of these would be phased in as running changes. Others would await the advent of a new E-type "series" some years hence.

Not many cars were made before the original "bonnet" was changed in two respects. Initially, the louvers atop that vast expanse were contained in a pair of add-on pieces; they soon were pierced directly into the sheetmetal instead. Also on the first cars, the hood couldn't be latched or unlatched without a special T-shaped tool inserted on each side; this cumbersome arrangement soon was replaced with cockpit levers. (Of course, those earliest, clumsiest E-types now command special prices from collectors!)

Over the next three-and-a-half years, up through autumn 1964, the 3.8-liter E-type was treated to similar rethinking in almost every area. Rear-axle ratio was raised from 3.31 to 3.07:1 for more relaxed high-speed cruising, though the original gearset would later return to recover the lost low-end acceleration. A permanent change was new piston rings that reduced oil consumption. They increased internal friction and cost a little power, but were somewhat offset by a thermostatically controlled electric radiator fan that also was quieter than the early constant-drive unit.

Still on the mechanical front, a new brake-operating system was adopted, allowing more consistent stops with less effort, and rain shields were added to the inboard rear discs. The handbrake received a self-adjusting mechanism so owners no longer had to do that dirty job by hand, and a more gentle clutch was installed.

Inside, noise was further reduced via added insulation and better door and hatch sealing. Heating and ventilation systems were upgraded, but the E-type cockpit still drew criticism for excess heat coming through the transmission tunnel, under which the exhaust pipes ran. The coupe's rear window got a heater of its own, an electric-wire type, to combat fogging. Seats were improved both in comfort and in length of fore/aft adjustment. A shallow well let into the floorpan opened up a little more foot room. Pedals were revised a bit, too, to meet complaints of awkward positioning.

Outside, a standard backup light sprouted above the twin exhaust pipes. At the front, the original plastic streamlining covers over the recessed headlights were replaced with ones made of toughened glass.

All these changes made a great road car even greater. Meanwhile, the E-types eagerly snapped up by racers were undergoing revisions of a different sort. Thanks to the factory's helpful attitude toward motorsports—and its stock of D-type racing components—an E-type committed to the circuit wars could wind up a pretty dramatic machine.

*Autosport* borrowed one such early "modified," a drophead, for evaluation on both road and track in late 1962. "The acceleration must be described as almost incredible," wrote Paddy McNally. Despite "far from ideal" conditions, and not thrashing the car out of respect to its owner, he was able to hit 60 mph in 5.2 seconds and tear past the quarter-mile pole in 13.3 at 108. Yet McNally felt sure that he hadn't seen all the speed the car had to give.

To get that much out of an E-type, one had to do a lot of work under the skin—perhaps starting with the skin itself. This particular car had been considerably lightened, to the extent of having its hood, doors, trunklid, seats and even its bumpers remanufactured in aluminum. The side-window glass was replaced with plastic, and the winding mechanisms removed.

Underneath, the original brake system had been revised with smaller-diameter but thicker "competition" discs, plus racing pads and a different servo—from a Ford. The suspension was retuned by lowering ride height, swapping the stock shocks for adjustable racing units, and cocking more negative camber into the rear wheels. While the back end was apart, the suspension carrier's stock rubber mounts were replaced with metal blocks "to prevent axle twist and also to cut out rear-wheel steering." Although the steering system itself was not mentioned in this case, some E-racers discovered they could make the car respond more crisply by removing the rubber rack mounts as well. All these changes would make the car too harsh on the street, but would make for

*Weathering new laws, expectations, and attitudes, the six-cylinder E-type gradually took on more amenities, a larger-displacement engine, even rear seats and an automatic gearbox. It grew safer and more reliable, but by the end of the sixties some pined for the irreclaimable wilds of its youth. If it was dimmed, it had enjoyed glory aplenty along the way, and much splendor was ahead. And as this '68 example shows, it suffered no privation of panache.*

all the right moves on the track.

The real centerpiece of this particular banquet was the engine. Atop its stock, albeit blueprinted, block was a D-type cylinder head complete with 10.0:1 compression, enlarged ports, a trio of two-barrel Weber carburetors, and a racing exhaust system. All this served up an estimated 300 horsepower. It carried through a lightweight steel flywheel and beefy competition clutch to closer-ratio gears installed in a stock transmission case. Final-drive gearing was variable, depending on venue; a 3.77:1 cog appears to have been fitted for the *Autosport* test, because the observed top speed of 128 mph was reached at the requested rpm limit of 6000.

"On the road," McNally reported, "this car proved extremely tractable, the engine never being temperamental or oiling up—throughout the period of testing the plugs were never touched. The engine pulled well at all revs and didn't just have top-end performance. Maximum power was found between 3000 [and] 5500 r.p.m., and the close-ratio gearbox allowed the driver to keep within this rev-band. Fast take-offs were helped by a Powr-Lok differential and the fixed rear-carrier, these two coping with take-offs at 4000 r.p.m., providing the driver was capable of holding the car in a straight line."

On the sensible side, McNally observed that this was "not a cheap car to run, at 10 m.p.g., but it certainly gave value for money." Well, cars like this tend to distort one's values; McNally actually thought the oil consumption "was quite low at one pint per 100 miles."

*Autosport*'s tester also felt the modified brakes "very good indeed with medium pedal pressure, and [they] never locked up, an amazing improvement over standard." He went on to praise the steering, which though heavy at low speeds was "superlative" when "really motoring." The clutch was very heavy, too, but smooth. Although ride was "firm, to the point of being hard," that minimized pitch and roll; overall, he lauded the handling as "ideal" and "superb."

On both road and track, McNally waxed on, "the road-holding was of a very high order. The tendency to understeer, with such power available, was no problem, and power-induced oversteer could easily be brought about. . . ." Rounding the famed Silverstone circuit, he found "the car could be made to do more or less anything, being extremely manageable and, needless to say . . . the car was controlled on the throttle."

This car, built and first raced by Ken Baker, went on to win numerous events and championships on the local level. Similar privately built racing E-types were soon numbered in the scores.

The ultimate harvest of all this hot-rodding was the dozen or so "Lightweight" E-types produced by Jaguar itself during 1963-64. These were mostly roadsters built up on specially made aluminum monocoques and

clad with aluminum body panels. Even the hardtop, fiberglass on the standard car, was of aluminum. The front space-frame remained steel, as on later D-types, and for the same reason: For this application, steel was effectively as light as magnesium and much easier to fabricate. But just about everything else was redone to save weight. The engines even revived the aluminum-block idea of the late 1950s, for a reduction of 84 pounds. Total weight-saving over the standard roadster was on the order of 500 pounds—some 20 percent. About 80 pounds of that was later put back on by installing a very stoutly constructed five-speed ZF gearbox from Germany.

While the Lightweights' suspension looked standard, it was, in fact, heavily modified, with different chassis pickup points for altered geometry, and certain parts adopted from contemporary Jaguar sedans because they were stronger. Torsion bars, springs, shocks, and sway bars were all stiffer, of course, as were the rear subframe mounting rubbers (the car's own designers evidently feeling some small amount of compliance was necessary). Front brakes were beefier, again thanks in part to some sedan pieces, while the rear binders were basically stock but newly fed with cooling air through special ducts. Instead of the normal wire wheels, the Lightweights had lightweight pierced discs that looked like the old D-type's but were 15 inches in diameter instead of 16.

Though all-aluminum, the engine also strongly resembled the old D-type's, having the wide-angle 35/40 cylinder head, fuel injection, and dry-sump oiling. Compression ratios ranged up to 10.0:1 and, during the engine's development life, horsepower reached as high as 344 at 6500 rpm.

That ultimate engine was installed in the ultimate E-type bodyshell, one of a series of super-streamlined fastback coupes wind-tunnel designed by Malcolm Sayer, Jaguar's astute, always active aerodynamicist. As entered at Le Mans 1964, this version was reported to whistle up as much as 6300 rpm in top gear down the Mulsanne Straight, which gearing charts showed to be 176 mph. By this time, though, Le Mans was being dominated by a different class of purpose-built racers capable of another 25 mph or more. Anyway, neither of the Lightweights that started that year's 24-Hours lasted to the finish.

To Jaguar enthusiasts who are also racing fans, the Lightweights were the high-water mark of the factory's competition activities in the sixties. But as modified road cars, they simply had built-in limitations and were soon outclassed in top-line international racing. Ferrari kept making faster coupes, and also stayed abreast of advancing technology in racing as a whole. Jaguar did not, or rather, did not seem to.

We know now that, in deepest secrecy, chief engineer William Heynes and his staff built and tested a state-of-the-art mid-engine racing sports car with a new 5.0-

*E-type glory shown with the factory "Lightweight" cars of 1963-64. They had aluminum bodies, reworked suspension, and an all-aluminum six of up to 344 bhp.* Opposite*: one of three that ran unsuccessfully at Le Mans in '63.* Above*: Malcolm Sayer's aero coupes hit 176 on the Mulsanne in '64, but like the Lightweights, were out-muscled by more advanced competition prototypes.*

liter, quadruple-camshaft V-12 of enormous power (reportedly over 500 bhp). Completed around 1966, this XJ13 could have carried the British colors at Le Mans against the Italian Ferrari "Prototypes" and the Anglo-American Ford GT40. But Jaguar management decided against the venture, considering it more important to concentrate limited resources on improving production cars.

Which brings us to October of 1964, when a 4.2-liter version of the 16-year-old XK six was introduced in the E-type. The important change was in the cylinder block, which had been recast to allow a bore dimension of 92.07 mm, 5.07 larger than before. To get that much of an increase involved something of a lash-up. Jaguar actually shifted the casting patterns to move the cylinders in relation to each other. Because the original engine had extra spacing between the middle two cylinders, numbers three and four, these were moved closer together; those at either end of the block, one and six, were moved farther apart. Overall block length was unchanged. Oddly enough, the existing cylinder head worked just fine; the slight misaligning of four bores and their combustion chambers didn't seem to trouble either the engine or its makers.

The crankshaft itself was redesigned, but retained the original 106-mm stroke, so the new displacement was precisely 4234.3 cc (258.4 cubic inches). This 12-percent increase in pumping capacity gave an 8.8-percent increase in torque, from 260 pounds/feet to 283, still peaking at 4000 rpm. Maximum horsepower held at 265 but was developed at 100 fewer rpm, 5400. Curiously, that was 400 rpm into the new, more conservative red zone marked on the 4.2 car's tachometer.

Accompanying this engine were numerous other improvements, or at least changes. There were new rings on the new pistons to further cut oil consumption, a new starter motor, a radiator core changed from aluminum to copper, and an exhaust system newly "aluminized" for longer life. A modern alternator ousted the old-fashioned generator, thus eliminating the gradual running-down of the battery in certain conditions, and the entire electrical system was switched to negative-ground. Another new brake servo was installed. Thanks to new seals, the lube interval for the front ball joints was upped from 2500 to fully 12,000 miles. At the rear, the suspension was altered slightly to prevent bottoming on severe bumps. An external SU fuel pump supplanted the original in-tank Lucas unit. Replaceable-bulb headlights gave way to sealed-beam units (a concession to the American market, one suspects), though these remained "under glass" for the time being.

Nor was the cockpit overlooked. Seats were improved again, armrests were added to the doors, a between-seats "cubby box" was put on the transmission tunnel for stowing bric-a-brac, and dash trim went from high- to low-gloss (again in deference to the U.S.). Exter-

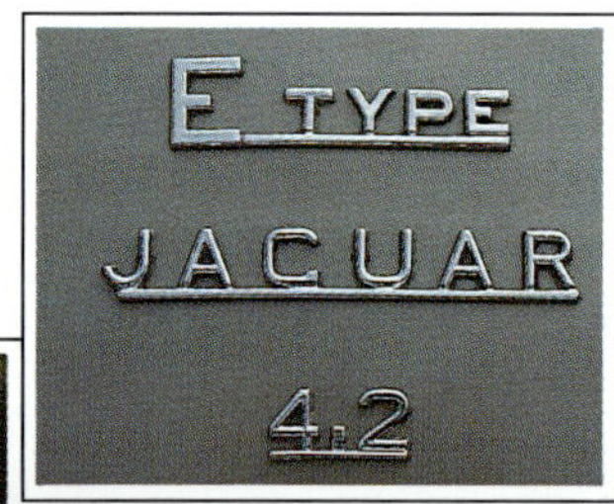

*A 4.2-liter six replaced the 3.8 in '64 and was followed by subtle style changes. Gone were the headlight covers, added was a back-up lamp.*

nal identification was far more subtle: just a small "4.2-liter" badge and covered hinges on the trunklid/rear hatch.

But of all the ameliorations made for the 4.2 E-type, the one most acclaimed was an all-new gearbox with—at last—synchromesh on all forward gears. In conjunction with a new diaphragm-type clutch, the new four-speed was as nice to use as the old one was unpleasant.

If the 4.2 was hard to distinguish visually from the 3.8—which remained in parallel production for a while—driving it is a surprisingly different experience. Where the 3.8 feels and sounds like a traditional sports-car engine, somewhat raucous and becoming especially vigorous in the upper half of its rpm band, the 4.2 hardly seems like a piston engine at all. Above slow-traffic revs, the familiar Jaguarish growl is replaced by more of a whine reminiscent of a gas turbine. And the 3.8's exciting shove-in-the-spine power step at about 3500 rpm is not felt in the 4.2, which doesn't seem or sound that eager to rev. The power peak may be 5400 on the dyno, but anything above 4000 seems superfluous on the road.

In the way "buff book" writers have of describing something new and fresh, initial reviews of the 4.2 E-type were generally favorable. "The biggest improvement," stated *Motor* in October 1964, "is the all-new, all-synchromesh gearbox. Gone is the tough, unrefined box that had accumulated a certain notoriety, in favour of one that will undoubtedly establish a correspondingly high reputation: although the lever movement is still quite long, it is fairly light and very quick, the synchromesh being unbeatable without being too obstructive."

Enjoyment was also found in the new engine's bottom-end strength. "Low-speed torque and flexibility are so good," *Motor* observed, that you can actually start in top gear, despite a 3.07:1 axle ratio giving 24.4 m.p.h. per 1000 r.p.m. Driving around town, this fascinating tractability can be fully exploited by starting in first or second and then dropping into top which, even below 30 m.p.h., is sufficiently lively to out-accelerate a lot of cars." Such tactics did tend to soot up the spark plugs, the magazine admitted, but a little high-rpm work would usually clear them.

When driven as a sports car rather than, say, as a limo, the 4.2's greater torque easily dealt with its slightly greater weight and somewhat taller final gearing. From a standing start, *Motor*'s test coupe reached 60 mph in 7.0 seconds, compared to the 7.1 the magazine had clocked with an early roadster three-and-a-half years before. The time over the quarter-mile was 14.9, versus 15.0. On the top end, the closed 4.2 was only slightly faster than the open 3.8—exactly 150 mph, versus 149.1—this despite the slinky fastback's much superior aerodynamics.

To achieve that velocity, by the way, *Motor*'s test crew apparently thought nothing of violating the

*The 4.2 at first retained the 265 bhp of the 3.8, but by '67 some were down from three carbs to two and had 246 bhp.*

tachometer redline by 1100 rpm. "All this performance is accompanied by astonishingly little fuss, the engine remaining smooth and mechanically quiet at all times . . . Even 6,100 r.p.m.—corresponding to 150 m.p.h.—does not sound unduly strained."

*Autosport's* John Bolster confessed to the same transgression in reporting an observed maximum of 152.5 mph: "The needle of the rev-counter has by then invaded the red section of the dial, but the engine is just as smooth as in the medium speed range, and that means very smooth indeed." His 0-60 time was 7.4, rather in arrears of *Motor*'s result, but his quarter-mile time was exactly the same: 14.9 seconds.

One cannot leave this point about high engine speeds without wondering whether it may help substantiate something long and widely suspected about published Jaguar performance figures—namely, that the cars loaned out for road testing were specially tuned for the task. Several marque historians have noted that private owners of production-line cars seemed unable to duplicate the press numbers—unless they spent a few hundred extra Pounds with the factory to have special engine work carried out. These modifications might well have raised an engine's willingness to rev, but also its oil consumption, its tendency to knock on normal gas, and so forth.

Such suspicions have cropped up in regards to other automakers. Some have even been proven. In this case, all that can really be said is that few, if any, private owners ever thought their own 4.2-liter E-types were near as fast as the publicity had led them to expect, nor even as fast as the 3.8.

That noted, it also must be said that the last tenth of a second and ultimate few miles-per-hour have very little relevance in real-world driving. Reaching the limits takes too much out of car and driver both. Besides, few E-type owners go about their daily rounds burning up clutches and grinding tires into nothingness. (Truer than ever now.) Few, frankly, ever cared to find out if their car was really capable of breaking 150 mph. Just as Jaguar built it, the E-type was fast enough to make a very quick, very enjoyable job out of any journey, and in itself was a possession that gave great pleasure. Surely that was enough. Numbers are interesting, but they aren't everything. The quality of performance is at least as important as its quantity.

As *Motor*'s test put it, "Preconceived ideas about speed and safety are apt to be shattered by E-type performance. True, very few owners will ever see 150 m.p.h. on the speedometer, but, as on any other car, cruising speed and acceleration are closely related to the maximum and it is these that lop not just seconds or minutes, but half hours and more, off journey times. Our drivers invariably arrived early in the E-type and the absurd ease with which 100 m.p.h. can be exceeded on a quarter mile

*If some 3.8 growl was gone, the 4.2 E-type's power was more accessible, and at 7.0 seconds 0-60 and a top end of 150, it was no slower. European-market cars kept classic knock-off wheel hubs—and a mallet to knock them off. Oil use was down and driveability improved, though Lucas electrics remained.*

straight never failed to astonish them: nor did the tremendous punch in second gear which would fling the car past slower vehicles using gaps that would be prohibitively small for other traffic."

This point was also touched by *Autosport*'s Gregor Grant in a piece he called "The Magic of an E-type." After a fast Continental trip in a 3.8 coupe, he wrote: "As a high performance touring car there are few machines to equal the E-type, and none at all in its price bracket. Effortless is the correct word to describe it, for it is a real mile-eater and also one of the least-fatiguing cars to drive . . . Driven intelligently, it is easily one of the safest vehicles on the road: it has superb steering and road-holding, the brakes on the latest version are as smooth and powerful as one could wish, and the acceleration is so vivid that overtaking can be done with the greatest confidence."

Bolster's test of the 4.2 also addressed passing performance: "The E-type is fundamentally a safe car because it is on the wrong side of the road for such a short period when overtaking." He also lauded, once again, the stability and comfort given by the all-independent suspension, the powerful disc brakes, the "very restful" silence as its aerodynamic shape slipped through the air; for Bolster, the E-type was "a superb car, a veritable magic carpet, which can make haste unobtrusively and automatically achieves fantastic averages."

Such was the real value for the average owner of the E-type Jaguar's competition breeding. An athletic automobile, like a physically fit body, simply handles everyday situations more competently. Let us sigh for the days and the places where the full prowess of such "magic carpet" cars could be freely and innocently enjoyed.

Even in lands where that was impossible, the E-type was popular. In fact, the United States was consistently its best market. Maybe Americans were mature enough to enjoy the marque's other virtues without indulging in its great performance potential. More likely, there was some "forbidden fruit" fantasizing going on. Anyway, in a 1965 test of the 4.2 "XK-E," as the E-type was generally known Stateside, *Car and Driver* focused less on speed than on sex-appeal: "There's something so sensual, so elemental in the appeal of that car that few men can resist its siren song. It's like that woman you used to love, the one you'd never waste another minute on. You can avoid her for months, but one night she calls. . . ."

Getting down to more tangible aspects, *C/D* found that the 3.54:1 rear axle specified for the U.S. gave 21.5 mph per 1000 rpm, which limited top speed to an "estimated" 130 mph (6047 rpm). But acceleration was commensurately fiercer, with the 0-60 time cut to 6.5 seconds. Over the quarter-mile, the results were pretty much as seen in Europe: 15.0 seconds at 98 mph. *C/D*'s test car

was a roadster with a dry weight of 2465 pounds. That rose to 2515 as "curb weight" with fuel aboard, and to 2800 as tested, the last distributed 49/51 percent front/rear. Gas mileage ranged from 16 to 22 mpg. List price in 1965 was $5525—not cheap (some $1500 dearer than an open Corvette, for example) but hardly "exoticar" territory.

Though *C/D* rated bumper protection and front seating room as Poor, it judged most points Excellent, Very Good, and Good. "Jaguar has been accused of ignoring its owners' anguished pleas in the past," said the report, "but this time they listened intently and fixed virtually everything [with the 4.2]." The magazine especially appreciated the new gearbox, saying it "can be banged into first at 40 mph without a qualm, although it's sometimes a little sticky when selecting the same gear at rest." Some other tests had mentioned that, some had not; perhaps a quirk in preparation of individual cars. As did almost every publication, this one marveled at the 4.2 engine's flexible torque delivery, a "locomotive-like ability to pull smoothly away from anything over 500 rpm in any gear." It also welcomed a reduction in brake and clutch effort, although some other testers complained the latter was still too stiff.

*C/D*'s writer was among those who found seating to be cramped. He acknowledged that the only way to provide more legroom would be to lengthen the wheelbase. "However, people of short to medium stature have been known to reach new heights of joy in the XK-E, and have accomplished 500-mile journeys in a single day's driving stint without discomfort. Limited luggage capacity in the XK-E roadster makes this sort of travel problematical, but it's noteworthy that the car makes you want to do it."

Expanding on his overall impressions, the writer added, "Driving the 4.2-liter XK-E is little different from driving its 3.8-liter predecessor. The driver sits proudly behind the same comprehensive—and comprehensible—instrument panel and bends the eager beast to his will . . . The short shift lever is just about where you'd put it yourself, and the throws are short, quick, and accurate. The steering wheel is placed at a nice angle for those who like to affect the Stirling Moss-Hero Driver style, and the steering is amazingly light for such a big car. It goes where it's pointed without fuss or surprises, and the handling is the kind that forgives the most ham-fisted cretin. The ride is sedan-like, and although the car isn't small the driver soon loses any apprehensions he might have had about that long nose and where it's going."

In all, *C/D* was "really very impressed by all the improvements that have been made to the XK-E, but we must be completely honest and admit that the things that really get to us are the looks and the noise. It's a Jaguar. It reeks of purest automotive erotica."

*Road & Track* made the same point with memorable simplicity. The E-type, it said, was "the greatest

*Jaguar unveiled the 2+2 E-type in '66. Wheelbase was up nine inches, weight by 200 pounds. Longer doors aided entry and exit and automatic transmission now was optional. This '70 example shows the Series 2's larger air intake, below-bumper parking/signal lamps, dual windshield wipers, and more widely spaced exhaust tips.*

crumpet collector known to man."

Ah, but having collected one's "crumpet," what then? Did the very prowess of the E-type as a courting car not work against it in later life? Jaguar had that covered, too. In March of 1966, it finally introduced a 2+2 coupe with enough extra interior room for a pair of occasional seats in back. One *Motor* scribe immediately zeroed-in on the real significance of this development: ". . . [I]t effectively turns the E-type into a family grand tourer and will therefore extend Dad's youth for at least another seven years."

As suggested by *C/D*, the 2+2 arrived on a longer wheelbase, up nine inches to 105 (three more than the old XK's). Created by stretching the sheetmetal center of the monocoque, it allowed correspondingly longer doors for easier access. In an attempt to make the overall profile look right, both roof and windshield were raised two inches. Some people actually thought this an improvement.

Some also felt that the longer, heavier 2+2 rode and even handled better than the two-seaters, and that its wider turning circle (up from 37 feet to about 42) was not much of a problem. Of course the added weight—roughly 200 pounds—and the taller body cut performance, but many people didn't think that mattered much, either. The "family Jaguar" was still plenty fast, and it was still a Jaguar. Plus, the factory used the extra driveline length to incorporate an optional item that had been popular on the old XK 150: automatic transmission. Thus, the 2+2 in every way diluted the sports-car concept still defended by its two-seat sisters. But it sold well.

Darker clouds were looming for the original E-type than a change in ownership base from younger singles to older marrieds. The auto industry and society as a whole were changing, too. By the second half of 1966, Sir William Lyons had established a relationship with British Motor Corporation, and within two years his once-independent Jaguar was merely a part of the new British Leyland conglomerate (after an interim BMC/Jaguar partnership as British Motor Holdings). Across the Atlantic, meanwhile, his largest market was busily redefining its very conception of the automobile and its place in society.

Jaguar met these realities in two steps during late 1967 and 1968. First came an interim model that enthusiasts now refer to as the "Series 1½," although the factory never used this; it hadn't yet gotten as far as a Series 2 at that point. The most immediately noticeable change was to the headlights, which lost their streamlined covers and were moved three inches forward to meet U.S. government notions of proper illumination. To satisfy another newly enacted regulation, switchgear changed from stiletto toggles traditional on British cars to rockers less likely to puncture a person in an accident. Under the hood, U.S.-bound engines had only two Stromberg carburetors, instead of three SUs, and were otherwise seriously detuned to reduce pollution. The official power figure dropped from 265 to 246. As something of a badge of shame, the engine lost its gorgeous polished cam covers for a nondescript pair covered with ribs and finished in "crackle" black.

These and further changes were incorporated in the officially designated Series 2 E-type introduced in October 1968. This remained very much the same basic design in most respects, but showed the heavy hand of the U.S. government and reflected the car-buying public's taste for more and more luxury.

Quickest way to spot a Series 2 from the front was by its more gaping "mouth" air intake—fully two-thirds larger. This sad degradation of the sleek, aerodynamically efficient original nose was required because—hold your heart—some customers wanted air conditioning. Probably the same people who had demanded the optional power steering that was now available.

Other Series 2 clues were larger front and rear parking/turn signal lights mounted below (instead of above) the bumpers. The bumpers themselves were now stouter and thus stronger, per Federal decree. Washington's mandated marker lights appeared on the sides, while twin backup lamps replaced the previous single unit. A new rear license-plate holder forced separation of the twin exhaust pipes, which moved farther apart and outboard. Windshields on the 2+2s got taller, so they now could use dual instead of triple wipers, and the glass was angled back an additional seven degrees, a boon to beauty. Cars sentenced to the increasingly restrictive American market also wore new "earless" nuts on their knock-off wire wheels—unless the customer had ordered the newly optional bolt-on disc wheels.

Alas, English and Continental cars weren't spared these changes, as Lyons had decreed that all Series 2s must conform to U.S. safety and smog specs.

But in many respects, what would be the last of the XK-engine E-types were the best yet, with better cooling, added comfort, a higher safety factor, and more. Still, their performance was very sad. A change in quoted horsepower ratings from British standard to European DIN made the new lower figures look worse than they were—down to as low as 171 bhp at a meager peak rpm of 4500. Actually, some authorities believe the once-mighty XK engine had been emasculated only 30 or 40 bhp.

The last cars were evidently so uninteresting to the motoring press that few tests were published. *Road & Track* and *Car and Driver* took their last looks in tests of then-new 1969 Series 2 cars. *R&T*'s was a coupe. It was priced at $6495 as tested, weighed 3018 pounds at the curb, and went 0-60 in 8.0 seconds. The quarter-mile clocking was 15.7 at 86 mph. Top speed was given as a

*In '69, this U.S.-market droptop cost $5858 and weighed 2750 pounds. It was aging, but time had allowed Jaguar to "lovingly bring the XK-E to full bloom."*

mere 119 mph, which meant 5500 rpm on its 3.54 axle. Fuel mileage was 15.9 mpg.

What did *R&T* now think of the sporting Jaguar? Much less: "Exciting though the E-type was when it was introduced in 1961, time has made it rather dated, inside and out. The interior, though retaining that wonderful smell of leather and the aura of a cockpit with a million controls and dials, lacks the spaciousness and ergonomics of more recently designed cars. Entry and exit are somewhat awkward and restricted . . . The XK engine doesn't seem comfortable when it's driven hard." And so on, all adding up to the impression that the magic that had once excused a multitude of sins had faded. "We hear that a new engine is in the plans for the E-type this fall; probably the basic car will be with us for another year or two. As it stands, it's a pleasant car in normal everyday driving . . . we can't really say we didn't like it. But we do think Jaguar can do better—and will before long."

*C/D* retested a roadster that cost $5858, up $333 or nearly 17 percent from the one examined four years earlier. As with the *R&T* car, horsepower was 246 at 5500 (the data panel and text disagreed; we cite the data panel, which was confirmed by other sources). Torque was 263 at 3000. The car's curb weight was 2750 pounds (distributed 48.8/51.2), compared to 2515 for the magazine's 1965 convertible. Running the same 3.54 rear-end gearing, the '69 made it to 60 in 6.7 seconds (versus 6.5 for the '65) and through the quarter in 15.3 at 90 mph (15.0 at 98). Top speed was once again estimated, but agreed with *R&T*'s 119 (versus 130 for the '65). Gas consumption was 16-19 mpg, compared to 16-22 in '65.

Though *C/D* still liked the E-type, and also what it seemed to stand for, this '69 report conveyed a certain poorly veiled impatience. "All you Faithful Readers still holding your breath for the V-12 Jaguar XK-F may exhale. It simply isn't time yet . . . If the apparent complacency in Coventry tends to infuriate you, look at it this way. Time flies for everyone except the Jaguar people. For them it passes in stately review. In this evolutionary context, the years have merely served to give them time to lovingly bring the XK-E to full bloom. A tasteful touch of refinement here, an unostentatious modification there. It's rather like breeding a Derby winner, or a Chelsea Flower Show champion rhododendron. As they undoubtedly say at Jaguar, 'You simply can't rush these things, old boy. It takes time.'"

By this account, Jaguar had changed a great deal from its heady days of the 1950s and early sixties, when there seemed to be more going on at Browns Lane than the most eager fan could keep up with. But a "V-12 Jaguar XK-F"? Well, yes, there was such a thing in gestation, but it would take another four years for Jaguar to give it birth.

*With a 3.54:1 rear, the '69 U.S convertible ran 0-60 in 6.7 seconds and the quarter in 15.3. Credible numbers, though top speed was an uninspired 119. The cockpit was less roomy than that of some newer rivals, but how many were more desirable?*

CHAPTER XV

# E-TYPE SERIES 3 V-12: TWILIGHT OF A GOD

On paper in 1971, it was a fabulous idea. The once-supernatural E-type was 10 years old and had lost some of its magic. At 23, the once-mighty XK engine had lost some of its power to emission controls. But to recapture the horses, Jaguar was developing a much larger, all-new motor for its sedans. So why not introduce it in the sports car and see if some of the E-type magic could be recaptured, as well?

After all, the ploy had worked back in 1948, when the all-new XK-powered 120 had created an aura of per-

*Its classic inline-XK six enfeebled by emissions regulations, Jaguar in 1971 turned to V-12 power for the E-type. Among alterations accompanying the move were a larger radiator inlet screened by formal grillework and the addition of subtle wheel arches to clear wider tires. The signature hood bulge was retained, even though it wasn't needed to clear the new engine.*

formance prowess—and, not incidentally, some real-life service experience—by debuting a powerplant principally designed for the Mark VII sedan. Only this time, perhaps the aura of an all-new engine might return the favor and juice up the image of a sports car now all too familiar.

And so in March 1971, Jaguar unveiled its long-rumored "new" sports model. Contrary to logical expectations, it was not called F-type but E-type Series 3. But who cared? What mattered was that it was motivated by the anticipated, wondrously exotic V-12.

Unfortunately, this "V-12E," as some called it, would never really replicate in the early 1970s the epic era of the original E-type in the early 1960s. And that was not entirely Jaguar's fault.

Creating automobiles used to be the purest pleasure. In the very early days it was a scientific activity, the entire focus being on the basic task of getting the newfangled contraptions merely to run, let alone reliably. Once the mechanical side was in hand, a maker was free to pursue the thing as an artistic endeavor, building machines to please itself. That quickly merged into its becoming a commercial enterprise, making cars to please people so they would buy them. But it was still fun. Until the Feds moved in.

Regulations in 1971 were not as choking as they would become, but they had already strangled the six-cylinder E-type. Neither in performance nor in appearance was it the thrilling super sports machine of a decade before. The once-crisp, bright engine had been saddled with anti-smog paraphernalia, and the once-clean, marvelously simple body lines had been uglified with new lights and bumpers mandated by law.

Every automaker was having to spend more resources on dealing with newly aroused social consciousness than on traditional automotive engineering. Besides that, roadways were becoming ever more congested, regulated, and policed, thus depriving the keen driver of exercising his cherished pastime. Both cars and driving just weren't as much fun any more; there was, in fact, an anti-auto lobby gleefully predicting the demise of the private motor vehicle. Even companies such as Jaguar were seeing customers concerned more and more with creature comforts and driving ease, less and less with performance and handling—with sporty style over sporting substance.

It was against this darkening background that Jaguar unrolled the drawings of a V-12 engine that it had designed in the very different times of just a few years before. Actually, the thought of someday building a Jaguar V-12 dated back to the early XK years of the late forties and early fifties. The idea finally took substantive form in the mid sixties, when some of the firm's competition-oriented engineers secretly built and tested a sports-racing car that promised to again raise high the Jaguar banner at Le Mans. This was the XJ13, an open two-seater with much of the immortal D-type in its body and chassis, but powered by a massive 5.0-liter, four-camshaft, 500-horse V-12 mounted behind the cockpit in modern mid-ships configuration. The man primarily responsible for the engine—code-named XJ6, by the way—was Claude Baily, a member of the XK-six design team.

According to the original thinking, the quad-cam twelve would have been developed and proven in racing, then detuned for docility and longevity as a passenger-car powerplant. A sound plan, as Ferrari, Maserati, and others had shown. For reasons both political and financial, Jaguar abandoned the racing program, but the initial "competition" engine would serve as the conceptual and experiential basis for a new V-12. However, this second unit was never intended for the race track. It was envisaged strictly as road-car power, primarily for a future range of sedans.

Why a V-12, exactly? Engineer Wally Hassan once

explained it in an interview with *Motor*: "Jaguar have always tried to provide luxury at a reasonable cost. Our problem was how to make the most reliable engine with the power to do the job and a lot of torque and refinement as well. We chose a V-12 formation because it gives perfect balance and, as vibration spells noise, this means a quiet as well as a smooth engine. In addition, the three-plane crankshaft is known to be the best configuration from a torque point of view."

Hassan left unstated the equally important commercial consideration that, at the time, only Ferrari and Lamborghini offered this many cylinders in roadgoing automobiles, albeit very high-dollar exoticars. Jaguar could bring the romantic song of the twelve to a great many more people.

As finalized, Jaguar's production V-12 was a bulky, beefy thing, big enough to encompass an ultimate displacement of a good 7000 cc (427 cubic inches). However, to meet the power requirements immediately foreseen, the engineers chose a swept volume of 5343 cc (326 cid) through a bore and stroke of 90×70 mm (3.54×2.76 inches). The angle between the two banks of six cylinders was 60 degrees, traditional for a V-12 because it gives even crank throws and thus smoother running. Both the heads and the block were cast in aluminum alloy for light weight—a savings of 116 pounds over the same design in cast iron. That number was known exactly because, concerned about noise, the factory did cast one experimental block in iron. But when installed in a test car, it proved not significantly quieter than the aluminum engine.

Within the block were pistons running in wet iron liners. For strength and, again, smoothest running, the skirts of the crankcase extended well down around the crankshaft, which was made of forged steel and whose main bearings were secured by four-bolt cast-iron caps. To be sure of adequate crankshaft support, seven of those bearings were specified.

The prototype "XJ6" twelve had twin-cam heads in the classic mold of the XK six. But Hassan, a former Jaguar man who'd returned to the company after its purchase of Coventry-Climax, was not impressed with the Baily engine's power output. Having just spent some years building world championship-winning Formula One engines—tiny 1.5-liter V-8s producing around 200

*Jaguar's V-12 was all-aluminum and displaced 5.3 liters. Ratings varied, but U.S. versions likely had 250 silken horses at 6000 rpm.*

bhp (a startling 133 bhp/liter), Hassan thought that out of 5.0 liters one should have seen much more than 500. This was not an especially kind judgment, for Baily and his team really hadn't had much chance to develop their racer. Moreover, compared to the racing Coventry-Climax eight, each of the V-12's pistons had to shoulder almost two-and-a-quarter times as much work.

However, the quad-cam layout did create problems with bulk, weight and cost for passenger-car applications. It also made for a lot of trouble in the mounting of accessories necessary on a street car.

Investigating alternatives through a series of single-cylinder test engines, Hassan decided that the most suitable head design was one with a single camshaft and valves parallel to one another and to their cylinder axis. Because bore was large relative to displacement, there would be ample valve area for good breathing at the moderate rpm this engine would be turning in its everyday traffic duties. Twin cams and hemispherical combustion chambers simply weren't needed. Not incidentally, using single-cam heads rather than a pair of XK-type twin-cams saved about 44 pounds.

In fact, not even combustion chambers were needed. At least, not in the conventional sense. For reasons connected both with smog control and manufacturing ease, Jaguar's new V-12 cylinder heads were machined completely flat on the bottom, flush with the valve heads. Combustion chambers were formed by depressions cast into the piston crowns. Such a design, not invented by Jaguar but not common, is known as a Heron head. However, Jaguar's engineers took pains to explain that their version was not a true Heron, because the piston "bowls" were broader and shallower.

These dished pistons were designed to give a compression ratio of 9.0:1 to suit 97-octane gas. Although fuel injection was becoming familiar in the industry, the engine couldn't be made to meet smog levels with such a system, so Jaguar fell back on emission-control Zenith-Stromberg carbs. There were four of these mounted at the ends of long, over-the-top intake manifolds designed to boost low- and mid-range torque. On the ignition side, though, Jaguar did break new ground with the first production application of a Lucas electronic system called OPUS, originally created for racing cars.

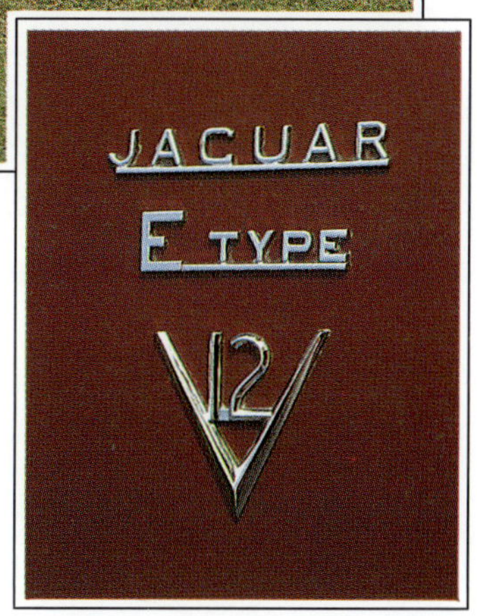


*The twelve was able to fit in the same engine bay as the XK six, but for better leg room, Jaguar shelved the convertible's 96-inch wheelbase and gave all Series 3 E-types the 105-inch span previously exclusive to the 2+2s.*

How much power did the big twelve really make? A precise number is a little hard to pin down. As finalized for the E-type Series 3, the power rating for European use was 272 bhp on the DIN scale at 5850 rpm. That would translate to 282 SAE. However, in U.S. marketing literature, the V-12 was said to deliver 314 bhp at 6200. The tachometer's redline was way up at 6500, by the way, and the engine could actually turn to 7840 before the valves started bouncing in their springs. Torque was listed in the same materials as 349 lbs/ft at 3600, but the European DIN measurement was 304 at 3800.

These differences can be explained by two factors. First, the American figures were idealized, being derived from dynamometer readings of a test engine unfettered by mufflers, air cleaner, fan, and other accessories. Second, importers like Jaguar could easily waffle about the potency of their "American versions" because, in 1971, the switch was on from advertising the traditional, but misleading, SAE-gross measurements to more realistic net outputs, and there was some public confusion as to what the conversion factors should be.

There was also this: U.S. emissions standards were then tightening so rapidly, and even inconsistently, that stated outputs could vary a lot from one model year to the next and, sometimes, even *within* a model year. This was particularly likely if a manufacturer needed to resort to different tuning to meet the stiffer emissions limits set by the smoggy state of California. We're inclined to believe *Road & Track*'s 1972 quotations as both plausible and definitive: 250 bhp at 6000 and 283 lbs/ft at 3500, both SAE net and applicable to all 50 states.

But none of this really mattered in the end. Few Jaguar buyers have ever based their purchasing decisions on the precision of engine outputs. The important point was that the Series 3 had more punch than the Series 2, enough to just about bring this much-heavier, much-less-aerodynamic car back to the performance levels of the original E-type. (More about aero and *avoirdupois* anon.)

According to measurements made for this book of display engines at the Jaguar factory museum, the V-12 measured 2 7/8 inches longer than the old inline-six, or 35 7/8 inches from rear of flywheel to nose of crankshaft. Height was actually a quarter-inch less, 26 1/2 from sump to the tops of the induction system; the engine itself, to the tops of the cam covers, was only 22 1/4 inches high. Of course, overall width was greater, especially when the V-12's sprawling induction system was considered. With this and all the other necessary ancillaries, the engine occupied a "box" measuring 44 × 39 × 27 inches.

So it was big—but not especially heavy. A good deal of arrant nonsense has been perpetrated over the years about the weights of Jaguar engines. Well, they're

*Where the old twin-cam 4.2 six cruised at 100 mph with little throttle opening, the single-overhead cam V-12* (above) *cantered at 120, and still had power enough to surge ahead when unreined. It was a new frame of reference.*

not made of fluff and foam, but they're not boat-anchors, either. In *The Jaguar Scrapbook*, marque historian Philip Porter published a list of weights of various Coventry powerplants "as installed on testbed." These were taken minus fan, air cleaner, clutch, and transmission, and without any fluids, but with flywheel, electrical equipment, and exhaust manifolds. In this trim, a 3.4-liter XK (for the 340 sedan, described in the next chapter) weighed 567 pounds, according to Porter. The 3.8, with its added cylinder liners, came to 592, while the 4.2, which had the redesigned block with different bore spacings, scaled 605.

And the 5.3-liter V-12? A mere seven pounds heftier, at 612. With the engine dressed for installation, the factory said its total weight came to 680 pounds. Granted, that was substantially heavier than a typical small-block Detroit V-8, but let's not forget that the Jaguar design was really a big-block enclosing a relatively small displacement—and was an inherently much-more-complex, more highly refined design to boot.

Fitting the V-12 into the E-type chassis took surprisingly little work. Lengthwise there was no problem; both Series 3 models rode the longer 2+2 wheelbase for reasons other than engine installation. However, the quartet of square-section tubes forming the main engine-bay space-frame had to be moved apart by several inches. Also, the top forward frame member was made detach-

*Series 3 steering was now power assisted and the downsized leather-wrapped wheel tilted and telescoped. But testers found little to like about the cabin, criticizing the ventilation system, control layout, and seats. Some also questioned the car's reliability and deemed the Series 3 a matter of "a magnificent engine in an outclassed body." But the Coventry cat still looked—and went—like little else on the road.*

*Jaguar fans anticipating an F-type got instead what looked like just a reworked E-type. But the silken new V-12, in harness with changes to suspension, a wider track, and more equipment and weight, effectively altered the character of the car far more than its looks suggested.*

*The ragtop had an aft parcel shelf in lieu of small back seats. It augmented the 4.75-cubic-foot boot.*

able for ease of engine installation; the powerplant was now dropped in from above whereas the twin-cam six-cylinder had been inserted from below. There also was some additional bracing and, because it had to be revised anyway to suit the altered frame dimensions, the firewall was beefed up.

While modifying the framework, Jaguar also re-engineered the front suspension, introducing anti-dive geometry to resist forward pitching under hard braking. Track dimensions were wider, the former 50-inch width at each end going up to 54.6 front, 53 rear. Some of that increase was due to the specification of wider, low-profile tires. Wheel travel was increased for a softer ride. Spring rates were revised upward, front brakes were given ventilated discs, and the rear discs were cooled by new air scoops. All these changes were necessary because the longer car with its added luxury equipment was heavier. With that and the wider tires, power rack-and-pinion steering was standardized. However, redesigned rubber mounts reduced the rack's former tendency to move sideways, thus cutting play in the steering. Rubber appeared

*". . . [A]ll we expected to try was a new engine. What in fact we drove was a new car—not a yowling, aggressive Ferrari-like (V-12) machine...but a very smooth, quiet and refined grand touring sports car." Jaguar grace once again.*

in torsion-bar and sway-bar mounts for the first time, again to soften the ride. Gas-charged shock absorbers, then in their infancy, were adopted to keep damping consistent as heat built up in the fluid with heavy use. These and numerous other detail changes all added up to a car that really was much newer than it looked.

Having wearied of complaints from taller members of the driving population who couldn't fit in the original E-type, Jaguar dropped the original 96-inch-wheelbase coupe and adapted the convertible bodyshell to the 105-inch span of the closed 2+2. Even shorter folk who might not need the extra nine inches of cockpit length could appreciate how this made for easier entry/exit and more useful reclining seats. Also more useful, thanks to the V-12's greater torque, was the optional Borg-Warner three-speed automatic transmission. The well-liked all-synchro manual four-speed remained standard, and was untouched save a larger clutch.

Any negative remarks about handling changes brought on by the long wheelbase were muted by the fact that Jaguar managed to alter the steering system to reduce the turning circle. It went from the very wide 42 feet of the 2+2 chassis to a much more manageable 36. Still, the sharper steering angles added nearly a full extra turn of the helm, from 2.6 up to 3.5 turns lock-to-lock. One might have expected a quicker ratio in view of the now-standard power assistance. However, the power assist did allow a smaller steering wheel, now 15 inches in diameter and leather-wrapped.

The high-roofed coupe bodywork remained virtually the same, but on the new convertible the rear occasional seats were exchanged for a small-but-useful luggage area. The drophead's doors were obviously longer than before and its windshield shaping somewhat different. On both models, wider tires required small flares for all wheel openings. Modern air-extractor vents appeared on both the coupe and the convertible's still optional lift-off hardtop to improve cabin ventilation, always an E-type sore point. Up front was a still-larger "mouth" to admit extra cooling air. The appearance was further aggravated by the first formal grille on an E-type. The broad cross-hatch dental work with a small vertical

*The coupe retained its rear +2 seating, as well as its high roofline, which could look a little ungainly from some angles.*

divider was topped by a Jaguar escutcheon, all rendered in sparkling chrome. Despite the enlarged opening, an additional inlet was added below. Both served a bigger radiator offering 40-percent greater cooling capacity.

Another body change was to the underside at the rear, where a lower line housed a larger gas tank. That addressed a point of criticism on the original E-type by bringing fuel capacity from 17 U.S. gallons to almost 22. Bumpers were refashioned at both ends, taillamps enlarged, and the standard steel-disc wheels redesigned with circumferential slots and nipple-like hubcaps. Inside was a noticeable lowering of the floor to improve foot room. This was partly dictated by a transmission tunnel made wider and stronger so as to force the engine safely down and away from occupants in case of a head-on crash.

Despite all these alterations, many of the Series 3's basic body stampings were interchangeable with those of earlier E-types. Even the hood bulge remained. It wasn't necessary to clear the new engine, but who would dare lose this distinctive and pleasing hallmark?

This emphasis on retaining the strong E-type identity raises some interesting questions. For example, is it possible that the Series 3 might have been better received had it not looked so much like the original? Or if had it been called "F-type," as widely expected? As long as it was spending so much money, Leyland-owned Jaguar might well have spent a little more on restyling (to better integrate the headlights, for example). Then the "V-12E" could have demanded acceptance on its own terms as something new. Instead, the first impression for some people was that the poor old E-type had somehow been made too long and heavy and soft; and besides, there was something wrong with its nose.

Nevertheless, the Series 3 really was a new car in most respects. "Perhaps a little naively," confessed Roger Bell in his April 3, 1971, *Motor* road test, "all we expected to try was a new engine. What in fact we drove was a new car—not a yowling, aggressive Ferrari-like machine with which, perhaps, most people associate a V12 engine, but a very smooth, quiet and refined grand touring sports car."

In several respects, Bell's prior expectations pro-

*Tightening emissions standards caused Jaguar to gradually reduce the V-12's compression ratio, while ugly bumperettes were added for '73 to meet U.S. crash rules. Most critics agreed that the E-type, while still appealing in its way, had been eclipsed in all-round roadworthiness by the new XJ12 sedan.*

duced "disappointment." That was the actual word, an extraordinary word to find in a British review of a high-end British car. "It is only at the top end of the rev range that you really begin to hear that beautifully distinctive and busy V12 purr," he observed. "At lower speeds exhaust noise is well subdued—perhaps too much so for sporting cars. Inevitably, there will be further disappointment that the car itself looks much the same as it did 10 years ago . . . We ourselves are disappointed that certain detail things have not been improved—such as the switchgear."

Bell did find several things to laud. The V-12, especially. "For its flexibility low down—by no means a weak point of the old XK [six]—and for its smoothness at the top of the rev band, the new engine is outstanding, altogether in a different class. It will pull strongly in top gear from under 500 rpm with an uncanny absence of vibration . . . Its ability to rev smoothly and willingly up to 6500 rpm is perhaps of greater significance, at least in the E-type, for the [4.2-liter] XK begins to feel rough at 4500 rpm. On twisty secondary roads where all good sports cars should excel, the engine's smoothness and eagerness encourage far more use of the lower gears than before."

Bell also liked the ride, "excellent by any standards," and the handling and roadholding; he called the general manners "impeccable." The power steering, though, was "rather too light" for his taste.

*Motor Sport*'s race reporter, the eminent Denis Jenkinson, was able to assess the new design on the basis of his own 150,000 high-speed miles in two examples of the six-cylinder model. He immediately disliked the driver's seat cushion, calling it too high and too flat for him. And he detested the automatic transmission fitted to one of the two Series 3s he sampled. Nor did "Jenks" approve of the way the big radiator opening had "a decorative bird-cage grille stuck up its nose." From a mechanical standpoint, he was surprised, after years of intimate experience with high-performance V-12 engines from other manufacturers, that Jaguar's was so smooth and silent—enough that he wasn't always sure it was running.

And at first, at low speeds, Jenkinson didn't seem

to see any advantage in having twice the number of pistons. "Throughout the whole afternoon I spent driving the two 12-cylinder 'E'-types, I found myself continually commenting that I would never know there was a V12 under the bonnet, especially when cruising about in a normal, leisurely 'seven-league-boot' fashion, but now and then there would be occasion to pull out and squirt past some traffic, and in the 70-110 m.p.h. range it really did come into its own. My reflexes and judgement being well attuned to 'E'-type performance in this speed range, I soon found that the V12 did not need anything like the time and space I was subconsciously allowing for overtaking, which made me realize how rapidly it was accelerating, making it an even safer and more long-legged car than the old six-cylinder."

At the end of his rapid afternoon, Jenks slipped back into his own Series 2 E-type, and ". . . as I motored off, my smooth, silent, silky, six-cylinder engine seemed as rough as the proverbial bear's hindquarters, and I realized that Jaguars [*sic*] have made an impressive step forward in refinement, which is so encouraging in these days of glorification of the cheap and shoddy."

On a later occasion, the same writer was able to try one of the new cars for a full week. He had been thinking of trading in his Series 2 for one. Why he decided against it is interesting. As Tommy Wisdom had noted after a drive in the original E-type a decade earlier, times change, and Jenks found they had changed once again.

"For all normal motoring purposes I could not see that the V12 engine gave any particular advantage over the six-cylinder apart from incredible smoothness and flexibility," Jenkinson observed in *Jaguar E Type*, a book about his experiences with the model. "Obviously it had a lot more power, and it did everything the 4.2 did but at 20 mph higher speed. Where the 4.2 would cruise at 100 mph with little or no throttle opening, the V12 cruised at 120 mph with the foot eased right back, but at the expense of 15 mpg against 21 mpg [12.5 versus 17.5 U.S.]. The acceleration of the 4.2 at 100 mph, for instant overtaking or getting ahead of an impending situation, was repeated by the

Car and Driver*'s '71 convertible test yielded these numbers: 0-60, 5.5 seconds; quarter-mile, 14.6 at 93; top speed, 135; price as sampled, $8809.*

V12 at 120 mph, though maximum speed was no better than [that of] the 3.8-litre E type. My eyesight, judgement and reflexes could not really cope with these increasing speeds as a continual way of motoring. I could see no justification for a V12 for my purposes, even forgetting laws and restrictions that were gathering fast in all directions." Those new limitations of the seventies, he sorrowed, had made Europe "no longer the happy care-free motoring paradise it had been" in the fifties and sixties.

Besides that, Jenkinson went on, the power of the new V-12 had outstripped the basic E-type chassis concept. Even with the Series 3's better tires, suspension, and brakes, "it was all too easy to run out of roadholding, steering and braking ability if you gave the 5.3 litres their freedom. The days of the E type were numbered . . ."

That sad fact was brought home forcefully a year later when Jenks made a trial run in a prototype XJ12, the running mate to Jaguar's new-generation XJ6 sedan powered by the selfsame 5.3-liter engine instead of the venerable XK six: "This . . . proved to be a giant of a car," Jenks wrote, "with road holding and handling up to using the full potential of the V12 engine. I soon realized that the E type era was over, for on a cross-country run you would have been hard pressed to have kept that big saloon in sight with an E type."

Even to Americans who had seldom experienced a "happy care-free motoring paradise," the Series 3 seemed outdated and inadequate. *Road & Track* subtitled its October 1972 drophead evaluation with this stark critique: "A magnificent engine in an outclassed body." What it found wrong with the latest E-type included these complaints: ". . . overall design lacks the sleek appeal of the original . . . ventilation system is antiquated and the controls laughable . . . no place for the driver to rest his left foot . . . clutch effort is quite high so [traffic driving] is more a chore than a pleasure . . . Seating is another area in which the Jaguar falls behind the times."

*R&T* also repeated that "reliability has never been a strong point of any Jaguar we have tested and this E-type was no exception." The actual troubles came to no

*Serious drivers shunned the optional three-speed automatic, but at least the twelve had the torque to handle it. Chrome V-12 trunk badging and the fab four exhaust tips identified a Series 3 E-type from behind.*

more than a worn alternator drive belt, but the magazine's own wariness of Jags in general paralleled that of its readers as revealed in surveys.

However, the *Road & Track* crew found much to like about the twelve-cylinder car: "Overall the E-Type is an easy car to drive and is most at home when driven hard and fast," they reported. The new leather-wrapped steering wheel earned good marks for its feel and adjustability—it had three inches of travel up and down its column and also retained the old E-type virtue of being adjustable for rake, although a wrench was still needed to take advantage of that. The brakes rated a "very good," the manual gearbox once again won high praise, and the ride comfort and the convertible's body rigidity on rough roads were both judged very good. Although the tires themselves didn't stick well at all—*R&T* said that the contemporary XJ6 sedan cornered faster on the skidpad despite being several hundred pounds heavier—Series 3 handling was found "neutral under all conditions except for extremely heavy applications of power. On such occasions the tail comes out gently and predictably."

There was quite a pronounced torque-veer, however: When the engine was pulling hard, the car swerved to the left, then darted back to the right when the driver's foot lifted. Some other contemporary tests mentioned the same phenomenon, by the way, but it should be noted that others made a point of saying their test cars showed no trace of such behavior; perhaps something to do with the condition of the rubber mounts in the rear suspension subframe, or in the way individual vehicles had been assembled.

As for the new engine, *R&T* pronounced it "a sheer delight, by itself almost worth the price of admission, and to some extent it atones for the sins of the outdated car. The V-12 is a lovely piece of machinery, lovely to listen to and lovely to behold. The exhaust has that hurried sound characteristic of a multiple-cylinder engine where the many explosions per revolution make it sound as if it's running faster than an engine with fewer cylinders. The idle is smooth and quiet with none of the mechanical busyness one normally experiences from the likes of a Ferrari or Lamborghini V-12. And the smoothness lingers throughout the rpm range . . . With the top down one becomes more aware of the nature of the beast lurking beneath that long bulging hood. Mechanical noise slips over and around the windshield and combines with the exhaust note to surround the occupants with sweet and sensuous sounds."

At the curb, *R&T*'s test drophead tickled the scales at 3380 pounds. That went up to 3680 "as tested" (driver and test equipment aboard), with a percentage distribution front-to-rear of 53/47. On the U.S. axle ratio of

*Bemoan the switchgear, criticize its aging lines, slam the seating, question its reliability, but the V-12E could inhale leagues of pavement at extra-legal speeds. And it sold reasonably well. By production's close in late '74, the Series 3 had tallied 21 percent of the 72,520 E-types built. Its last hurrah was as Group 44's SCCA B-Production champ in '75* (right).

3.54:1, the time taken to 60 mph was 7.4 seconds, the quarter-mile came up in 15.4 at 93 mph, and top speed was calculated to be 135 mph at 6000 rpm.

*Car and Driver*, testing a similar Series 3 convertible, confirmed that top speed exactly, though it was "observed." However, acceleration was far better: The magazine clocked the quarter-mile in 14.6 at 97 and 0-60 mph in a resounding 5.5 seconds.

After a drive right across the United States, some of it at high speeds in then-unlimited Nevada, *Road & Track*'s fuel consumption worked out to 14.5 mpg (*Car and Driver*'s ranged from 11 to 14). Base price of this 1972 open car was $7599, but air conditioning, wire wheels, and stereo raised the tab to $8809.

In summarizing the Series 3, *R&T* defended its numerous criticisms this way: "We've been harsh on the car, but we believe justifiably so. When the same manufacturer can produce such an outstanding car as the XJ6, it not only spoils us for anything he produces thereafter, but makes it exceedingly difficult for us to justify the existence of a car that is not as excellent."

Sadly, as part of the increasingly troubled British Leyland, Jaguar could only let the E-type slip further. Tightening U.S. emissions standards prompted compression to be eased as the years passed, and Washington's mandate that bumpers be able to survive five-mph shunts prompted ludicrous rubber-block bumperettes for 1973-74. By that time, only the convertible could find buyers, and demand for that was tailing off. Production was apparently shut down around the end of 1974, although Jaguar kept the fact quiet until early '75 so as not to complicate clearing the large stocks still on hand. The very last Series 3, an Open Two Seater appropriately finished in black, rolled down the Browns Lane assembly line and straight into the factory museum. It was the 15,287th Series 3, and the end of a proud E-type line that reached total production of 72,520.

Before concluding the generally dim V-12E story, it is pleasant to record one brief shining moment. "Camelot" in this case was the state of Virginia and the headquarters of a racing team called Group 44. Sponsored by Jaguar and headed by Bob Tullius and Brian Fuerstenau, it entered race-prepped V-12 roadsters in SCCA B-Production events during the summer of 1974—ironically, just as E-type production was winding down. Very competitive that first year, driver Tullius just missed the national title, but came back to win it at the end of 1975. It was a grand success in the glorious Jaguar tradition, but also one that reinforced the reputation of the crisis-wracked conglomerate that owned the marque: It came too late to do any good.

Jaguar's sun, it seemed, was sinking fast.

CHAPTER XVI

# THE SEDANS: JAGUARS ALL

Jaguar may have captivated the public's fancy with sports cars, but it captured their wallets with sedans. In round numbers, for each pure two-seater the company sold in the years it offered such a configuration, it attracted three buyers who required rear seats.

None of these purchasers seemed to feel short-changed by their longer vehicles. The sedans might not have had quite the same style and speed as an SS, XK, or E-type—what sedan could?—but they embodied every other Jaguar virtue. They were sporty, sometimes even rakish in their own distinctive ways and delivered generally outstanding road performance compared to their direct rivals. They also boasted interiors combining comfort and elegance in ways that made them very pleasant places to be and, in keeping with William Lyons' most cherished marketing precepts, were offered at prices that were always remarkably low for the value perceived. In short, the sedans were Jaguars all.

For many years, another of Lyons' principles seemed to be that his lineup should always include two

distinct sedan sizes. Thus, when he launched the dramatic-looking, six-cylinder SS1 for 1932, he sent out a similar-looking but physically smaller four-cylinder companion. Called SS2, it was first powered by an engine with a mere 1006 cc (just 61.4 cubic inches), then graduated to a 1052 (64.2 cid). Two years later, it grew up a little, both in length and displacement, with two engines offered concurrently: a 1343 cc (82 cid) and a 1608 (98 cid), both still side-valve designs. At various times, body styles included 2+2 coupes, full four-seater "saloons" and four-seat drophead tourers.

The SS2 was considered a very nice little car, with modest performance but decent road manners; some SS enthusiasts even preferred it over the more cumbersome SS1. Typically, it sold for two-thirds the price of its bigger sibling. Over a five-year life, through model-year 1936, total SS2 production reached very nearly 1800 units, compared to a little more than 4200 SS1s (this according to the late Andrew Whyte's definitive *Jaguar: The History of a Great British Car*).

When Lyons introduced his overhead-valve, 2.7-liter SS Jaguar sedan at the end of 1935, he took care to accompany it with a "One-and-a-Half," again very similar in appearance but slightly smaller. Only the side-valve 1608 was offered initially; it was then converted to an ohv powerplant of 1776 cc (108.4 cid) late in 1937, the same time the bigger car received its optional 3.5-liter engine. Through the war-shortened 1940 selling season, the "One-and-a-Half" SS Jaguar in both its engine packagings proved popular enough to garner almost 7300 sales, about 300 more than for the bigger sedan.

Both sedans and all three engines returned postwar, virtually unchanged from prewar form. But the new Jaguar Cars, Ltd., had not acquired the tooling for the little four-cylinder when it bought rights to the six from Standard. Thus, when the Mark V came along in the fall of 1948, it came only with a six.

At that stage, Lyons had plans for a four-cylinder engine of his own: a shorter, smaller-displacement version of his beautiful new twin-cam XK six. Experimental units of 1995 cc (122 cid) were built and extensively tested. One was even installed in a special speed-record streamliner, which the famed "Goldie" Gardner hurled to a record 176 mph at Jabbeke in Belgium in September 1948. The following month, still basking in the ensuing publicity, Jaguar announced a production version of this engine as an option for its swoopy new "Super Sports" roadster. The resulting "XK 100" was identical to the 3.4-liter, 160-horsepower XK 120 except, of course, for its twin-cam four, whose stated bhp was 95.

But as events unfolded, Jaguar had neither the manufacturing capacity nor the commercial need for a

*Sports cars put Jaguar's name in lights—particularly in America—but the sedans sold far better worldwide. Many of Coventry's saloons were quite adept performers, though, and cultured, too, with fine woods and leathers.* At left: *the Mark VII (1951-57), the second-generation XJ6 ('86-present), and the Mark X ('62-67).*

*Coventry's bread-and-butter cars in the 1930s were its range of 1.5-, 2.5-, and 3.5-liter SS Jaguar sedans. Note how the prewar styling of the '39 "Two-and-a-Half"* (above and right) *carried over to the '46 "One-and-a-Half"* (top).

secondary sports car. It was all the company could do to get the XK 120 and its big sedan stablemate, the Mark VII, into production. The same factors also precluded a new small sedan. Though the four-cylinder XK hung around for several years, both as a good idea and in the tangible form of some 50 completed engines, it would go no farther. When Jaguar finally turned to a new compact sedan in 1955, it used a short-stroke version of the XK six.

Meanwhile, the larger saloons were doing well in the market, and surprisingly well in sporting competitions. During its three-year production life, the pushrod-engine Mark V accounted for well over 10,000 sales while putting up an impressive record in long-distance rallies. Perhaps its most notable accomplishment was third overall in the 1951 Monte Carlo event, the rally world's equivalent to Le Mans.

Its successor, the Mark VII (they skipped a "Mark VI," remember), took up this good work right from its introduction in October 1950. Built on the Mark V chassis, with independent front suspension and live-axle rear end, it was a bulky-looking car, a full five-seater with a 120-inch wheelbase and curb weight of close to 4000 pounds. Yet the 160-bhp XK twin-cam could push it beyond 100 mph, and with its stiff chassis, supple suspension, and flowing body lines that obviously recalled those of the XK 120, the Mark VII offered genuine Jaguar performance and style to those who felt constrained to approach their sporty motoring in a more dignified manner.

The Mark VII was an immediate and lasting success. Even the Queen Mother chose to be chauffeured about in one. Stirling Moss chose to do his own chauffeuring, taking a race-prepared example to victory in a major

*Though it used another manufacturer's engine and chassis, Lyons' four-seat SS1 Saloon of 1933* (left) *had an unmistakable presence that helped set the tone for all the Jaguar sedans that would follow.*

touring-car event at Silverstone in 1952—which the model promptly won again over the next four successive years. And in 1956, a Mark VII won the Monte Carlo outright.

Late 1954 brought a replacement Mark VIIM with freestanding foglights and a boost in engine power to the same 190 of the concurrent XK 140 sports car. The VIIM continued into 1957, by which point sales of the first XK-engine sedans stood at exactly 30,200. But Jaguar had decided it was time for a replacement, though in view of that sales tally, it didn't dare replace much. So though improved and refined in several ways, the "new" Mark VIII was really just a VII with one-piece windshield, two-tone paint, and a power boost to 210. It found 6332 customers into 1959, when the Mark IX arrived to continue the upgrading theme. Featured were the 3.8-liter XK 150 engine with 220 bhp, plus four-wheel disc brakes. Sales totaled 10,005 through 1961.

By that time, the role of race-and-rally sedan had been assumed by the long-postponed postwar compact. Appearing in September 1955, this "2.4 Saloon" proved that the intervening years had been well spent. Its most important technical feature was "unibody," or combined body/chassis, construction. Jaguar engineers under William Heynes had been experimenting with such designs ever since the war years, when they built a couple of prototype military vehicles on the monocoque principle. Then came the prototype monocoque racer of 1953 that evolved into the milestone D-type a year later. That car, of course, won the 1955 Le Mans 24-hour race, just a few months before the new small sedan was unveiled.

The 2.4 Saloon was no racer, but it made good use of the unibody concept, with a stiff, lightweight structure that maximized the effectiveness of its engine, just as any race car must do. The 2.4 also represented a determined effort to make an aerodynamically clean bodyshell, combining Jaguar's then-characteristic ovoid grille and "center-weighted" headlight position with the round nose and tapering tail of a torpedo, a shape Lyons used to call "the rotund look."

Perhaps to emphasize the overall tapering-tail appearance, the rear track of the new compact was four inches narrower than the front: 50⅝ inches versus 54⅝. (Heynes was a Citroën admirer of long standing, but it is surely pure coincidence that the French maker's epochal DS-19 sedan, introduced at the same time, also had a narrower rear track). Though the 2.4's rear suspension was a live-axle type, as was still normal at Jaguar, it had something different in the springing. Instead of conventional semi-elliptic leaf springs, Jaguar reverted to old-fashioned quarter-elliptics, almost literally "half-springs" attached at their thickest points to the bodyshell and cantilevered rearward toward the axle. Lateral location was assured by a Panhard rod, a device that Jaguar had first used on the 1953 C-type racer.

Front suspension was an even greater change from Coventry practice, for there were coil springs for the first time, and the entire assembly was carried on a rubber-mounted subframe. The latter was dictated by the need for acceptable cabin isolation from the road noise and harshness that unibody construction tends to amplify.

The same quest for refinement had finally killed production prospects for the four-banger XK. The new compact thus used the smoother-running six, albeit a

*Jaguar's first fresh postwar model was the Mark V* (both pages). *Its cultured interior appointments and thoughtful amenities, such as the sunroof and hidden tool kit, carried on Lyons' tradition of value for money. And underneath was a new, extremely rigid chassis and a very capable independent front suspension, which gave the five-seater outstanding ride and handling.*

*With its roadworthy underpinnings and 2.7- and 3.5-liter sixes, the Mark V of '49-51 was a good seller and a popular rally car; it finished third in the '51 Monte Carlo event.*

"short-deck" version, with a bore and stroke of 83 by 76.5 mm for a displacement of 2483.5 cc (151.5 cid). Output was 112 bhp at 5750 rpm.

Riding a 107.3-inch wheelbase, the 2.4 Saloon was only a bit longer than the sporting XK 140 and, at approximately 3200 pounds, was little heavier. Here, then, was the very definition of the term "sports sedan," and it delighted Jaguar's public. So much so that this basic design would become the mainstay of Coventry's entire product line; with numerous revisions and additional engine options, more than 185,000 of the small sedans would be built over a lengthy production lifespan of 14 years.

The road to this landmark longevity was paved with the inevitable installation of larger XK engines. The 3.4-liter came first, in 1957. The 3.8 joined the range in 1960, making this sports sedan downright exciting, its 220 bhp giving an effortless 0-60-mph time of about 8.5 seconds and a top speed of 125 mph or more.

Coincident with adding the 3.8, Jaguar extensively revised the basic car into a Mark II ("Two") version. This was signaled by several bodywork revisions: thinner pillars, larger door and rear windows, chrome-plated window frames (including windshield), a heavier central grille bar, and similar detail changes. Inside, a new dashboard layout grouped primary instruments in front of the driver instead of in the middle. Underneath was a rear axle with a three-inch wider track, and various optional performance items were made available, including limited-slip differential, servo-assisted all-around Dunlop disc brakes, and wire wheels.

*The feline contour of the XK 120 sports car was clearly evident in the fender lines of the Mark VII saloon of '51-57. It's seen here in prototype body shell* (above) *and in finished form* (right). *This was a landmark sedan for the company.*

The 3.8 Mark II added many laurels of its own to the competition successes already scored by the "Mark I"s, and the compact Jaguar remained a common sight at the fronts of races and rallies literally all over the world. It even managed four consecutive victories in the Tour de France, that long-distance competition of many varied challenges so long a province of Ferrari's two-seat Grand Touring coupes.

The cars' performance also impressed the British police. Continuing a fondness for Lyons' products that had begun with Swallow sidecars in the 1920s, departments all over England bought small Jaguars by the dozens.

Continuing to mine the popularity of its compact four-door in 1962, Jaguar installed the nice little 2.5-liter, hemi-head V-8 it had acquired by purchasing Daimler. Special grillework was made up incorporating the old company's traditional "fluting" in the top of the surround, and furnishings were made more luxurious in keeping with Daimler's upper-crust image. Simply called the Daimler V-8, this variation was sold alongside the six-cylinder Jaguar models, including the 2.4, which remained in production with all the Mark II improvements.

That made four models on one basic platform, and Jaguar added two more in 1963. Named 3.4S and 3.8S—"S-type" for short—their main mechanical feature was the same independent rear suspension introduced three years earlier on the E-type. That increased wheelbase by a fraction of an inch, but overall length grew six inches, to 187, thanks to a more gracefully tapered tail (like that of the contemporary big Mark X sedan,

*The wonderful 3.4-liter XK six* (left) *that emboldened the XK 120 sports car was developed with the Mark VII in mind. It had a similar impact here, lending genuine muscle to a five-seat saloon that gave up nothing in cabin comfort or style.*

described a bit further on) that happily enclosed some additional luggage space. That was the S-type's most recognizable difference, although a modestly revised roofline gave a little more headroom at the rear, and the front end was slightly restyled to eliminate the chrome headlight rims in favor of small hoods like those an American hot rodder would call "frenched." Though the S-type was obviously a better car, with a superior ride for rear-seaters especially, it didn't sell well. The reason, most probably, was that its mismatched lines just didn't look right to most people, although, of course, the basic car was now pretty long in the technical tooth. The 3.4S would see 10,036 copies, the 3.8S 15,135, both through 1966.

The final evolution of this series occurred in 1967-68. First, the 2.4 Saloon was renamed Type 240 and the 3.4 became a 340. Both retained the original "short-tail" body, but the previous 3.8-liter version was dropped and the two S-types were replaced by a single 4.2-liter model logically called 420. This last retained a long S-type tail, but looked somewhat better balanced thanks to a new squared-up front with four headlights (again mimicking the big Mark X). On the Daimler side, the short-tail V-8 model was rechristened V8-250, and a fluted grille was applied to the 420 to make a new Daimler Sovereign, though this confusingly carried the Jaguar XK six. None of this "badge engineering" created additional sales acceptance (the 420, for example, found only 9800 buyers). By the end of 1969, Jaguar's compact sports-sedan family was history.

So, for most customers, were Jaguar's senior four-doors. In October 1961, seven months after the technically advanced E-type illuminated the sports-car heavens, its rear suspension concept appeared under a brand-new big sedan with unibody construction *à la* Mark II. Designated Mark X ("Ten"), it was physically the largest car Jaguar had ever attempted, spanning the familiar 120-inch wheelbase but, at 202 inches overall, half a foot longer than the Marks VII/VIII/IX. It was also proportionately wider: 76 inches. Well, Jaguar was still marketing its cars under the longtime slogan "Grace, Pace, Space." At least the Mark X had that last quality covered.

But not "pace." Weight was substantially over two tons, and even with the 3.8 engine, the Mark X didn't have much accelerative ability. Contemporary tests reported 0-60 times of just under 11 seconds or so, which wasn't really brisk enough anymore to impress anyone, especially in city traffic. Given a long enough straight, however, the XK could wind the car up to about 120. It was to address the standing-start "pace" problem that Jaguar developed the torquier 4.2-liter XK. This was duly installed late in 1964, and improved 0-60 acceleration by about half a second; top speed went up a couple of mph. Two years later, the Mark X was given very minor cosmetic changes and redesignated 420G. ("G" for giant?)

Though this biggest Jaguar was never an enthusiast's car, many people became enthusiastic about its opulent elegance, its smooth comfort and, no doubt, its impressive bulk. It was particularly well-liked by the sort of Londoners who employ chauffeurs to drive them to work every day. There was even a limousine version with "division," a glass partition between front and rear seats. Later on, in 1968, the Mark X platform formed the basis of the Jaguar-built Daimler Limousine. That regal carriage, the very traditional-looking, resolutely square-rigged con-

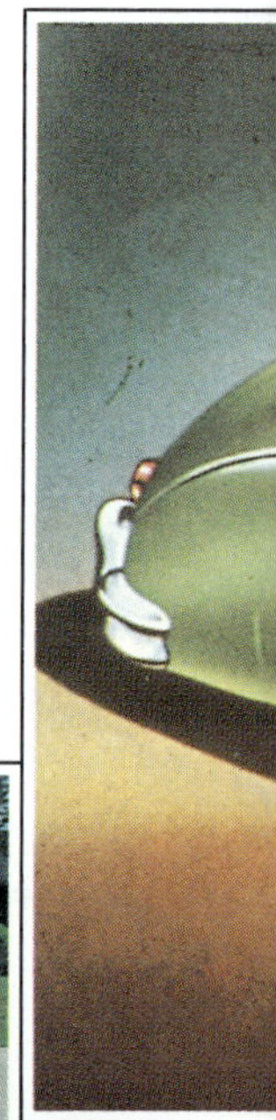

*Winning the '56 Monte Carlo rally was testament to the tenacity of the Mark VII* (above). *Evolution of this highly popular saloon line concluded with the Mark IX of '59-61* (right, both photos), *which boasted the 220-bhp 3.8-liter six from the XK 150 and four-wheel disc brakes.*

*In 1955, Jaguar returned to its prewar practice of offering saloons in two size classes by launching the "2.4 Saloon"* (top, both photos). *Its 107-inch wheelbase was 13 inches shorter than the Mark VII's and it also was notable as Jaguar's first unibody sedan. In '57 came a 3.4-liter version* (left) *and in '60, a 3.8. These defined "sports sedan."*

veyance officially titled DS420, was taken to heart by the sort of Britishers who are driven between the various royal palaces in and about London.

Altogether, the Mark X in its 3.8, 4.2, and 420G guises accounted for some 24,200 sales through its farewell model year of 1970. Exact breakouts are 12,999 Mark X 3.8s, 5680 Mark X 4.2s, and 5554 of the 420Gs. The Daimler Limousine, at something over 4000 made, continued in production at Browns Lane up through the time this book went to press. One steady customer was the Queen Mum, who had forsaken her Mark VII for a series of these huge, hand-crafted Daimlers. (Each time she changed cars, she made sure her favorite back-seat radio was transferred.) Thus was the high duty shouldered in its last automotive application by the grand XK Jaguar engine, now well into its fifth decade of service.

As for the volume range, Jaguar finally gave up on the old idea of fielding two sizes of "saloon." By the mid sixties, Sir William had concluded that an in-betweener could cover both bases—actually, the half-dozen bases covered rather extravagantly by the various iterations of the Mark II and Mark X. Accordingly, in 1964 he put into motion a prototype program called XJ4 (the letters evidently denoting "Xperimental Jaguar"). Four long years of development culminated with the September 1968 introduction of a new production sedan called XJ6. Not only would it soon sweep away all of Jaguar's existing four-doors, it would sweep buyers off their feet while setting new standards of performance for sports/luxury sedans.

Although the new sedan retained the hallowed XK engine, it did not bear those famous old initials because, right from the beginning, it was planned for a second powerplant. This would be Jaguar's long-rumored, super-secret V-12, which the XJ engine bay was designed to accommodate. When this variant finally did appear almost four years later, it was inevitably titled XJ12.

The new basket into which Coventry put so many of its commercial eggs arrived on a wheelbase of 108.8 inches and stretched to an overall length of 189.5. Though also retaining the unibody construction of its immediate predecessors, the XJ6 employed rubber-isolated subframes for both front *and* rear suspensions. The XK engine was again carried on the front structure to further separate occupants from nasty noise, vibration, and harshness.

Front suspension was another coil-spring arrangement, albeit with a broader track dimension of 58 inches. And there was a new wrinkle: "anti-dive geometry," to keep the nose from dipping under braking. At the rear was the familiar E-type independent-rear suspension, though modified to achieve a track 0.6 inch wider than the front (and 8.6 wider than the E-type's). Dunlop chipped in with new low-profile radial tires developed especially for this chassis. Another interesting feature was something carried over from the compact S-types: twin fuel tanks, one in each rear fender. The idea was a more usefully cubic luggage boot.

Primary power was predictably provided by the now-familiar 4.2-liter six, tuned for 240 bhp. For the first few years, however, there was also a 2.8-liter version for sale in countries like France where motor vehicle taxes

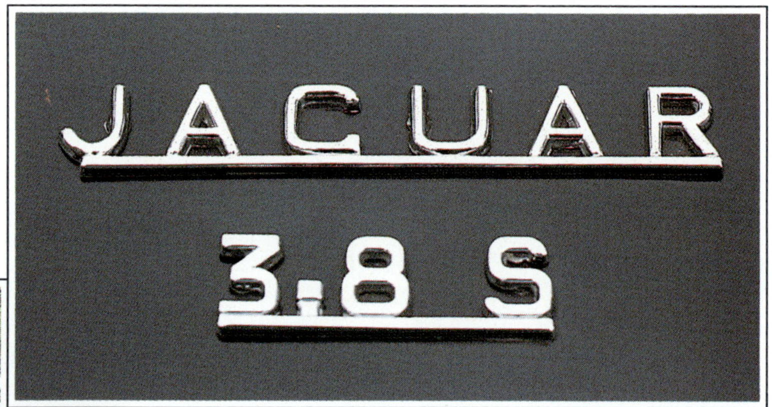


*Along with adding the 3.8 XK 150 engine, Jaguar extensively revised its mid-range sedans for 1960 and redubbed the line Mark II. Among the changes was a new instrument panel with gauges grouped before the driver instead of in the middle of the dash. The Mark II 3.8S version of '63* (this page and opposite, bottom) *got the E-type's rear independent suspension under an elongated tail.*

*William Lyons liked to develop his automotive designs from actual full-scale mockups, but drawings from the mid fifties* (above) *show Coventry also explored some sedan concepts on paper. Thankfully, none of these were built.*

rose sharply above that displacement. This shorter-stroke XK was achieved by combining the old 83-mm bore with an 86-mm stroke, giving swept volume of exactly 2791.9 cc; rated horsepower was 180. Unfortunately, this engine would prove to lack reliability, and would be quietly dropped in 1973. But the "big-engine" model, though not without its own problems, would turn out to be one of Jaguar's major success stories.

Generally considered the last body style for which Sir William was primarily responsible, the XJ6 was also widely acclaimed as one of his best-looking. Some Mark X/S-type elements were still evident, such as the four headlights of two different sizes under little "eyelids," and the pronounced "tumblehome" of the bodysides. A new square grille gave the touch of formality expected in a posh sedan, but was handsomely balanced by a subtle dip and rise in the rear fenders that evoked the sinewy cat-like strength of the voluptuous XK and E-type sports cars. Topped by a lowish, thin-pillar roofline with trim chrome moldings, the overall effect was totally new yet still Jaguar-elegant: "grace" with more than a suggestion of "pace"—and more space than in the old Mark II. Somehow, everything came together in a cohesive whole not evident since Sir William's first compact sedan a dozen years earlier. Styling was further refined for both aerodynamic drag and stability in extensive wind-tunnel tests.

Performance-tested in European trim, where it weighed about 3800 pounds empty, the initial 4.2-liter XJ6 could do 0-60 in less than nine seconds with manual transmission (the customary four-speed with optional electric overdrive) and reach nearly 125 mph. Moreover, it did so in such hushed refinement, and offered such a smooth ride allied to such sure-footed handling, that most every journalist declared it a milestone in the automaking art. Yet at the same time, they noted curious deficiencies, such as weak ventilation and heating and excessively powerful power steering. Many customers reported disappointments and troubles of various sorts, yet they still loved it. The XJ6 was a true Jaguar—unless it wore the alternative radiator shell, in which case it was a true Daimler (a new Sovereign).

And true to form, Jaguar continued to make it better. In 1972 came six- and 12-cylinder "L" models stretched four inches between wheel centers (to 112.8) to answer cries for better rear-seat legroom. The following year brought more substantial alterations, enough to justify a Series II ("Two") designation. Instant identification was provided by a revised nose with a shallower grille, this to accommodate a beefier, higher-set bumper necessary to meet 1972's new U.S. regulation for five-mph "crash protection." The dash layout was revised to meet other new Federal standards, and environmental systems were improved (notably the optional air conditioning).

Come 1975, Jaguar announced a brace of derivative two-door "C" models, the XJ6C and XJ12C. Built on the original sedan wheelbase, these had rounder, slightly abbreviated rooflines minus B-pillars, making them true hardtop coupes of the sort Detroit was then abandoning in fear of a pending U.S. rule that would outlaw such things, though it would never materialize. A trendy American-style vinyl roof cover was included, allegedly to hide some none-too-tidy hand finishing.

The XJC was a lovely answer to the pillarless Mercedes and BMWs of the day, but it never really caught on. That likely was due to restricted rear-seat space, the

*Silverstone, Monte Carlo, the Tour de France—just some of the venues conquered by Jaguar's sports saloons. Particularly popular was the 3.4-liter version, renamed the 340 in 1967* (both pages). *Along with the 240, it was among the last of a sedan family whose production reached 185,200 through 1969.*

*The 240/340 saloon returned to the popular "short-tail" body* (above) *that had been abandoned for the S-type. The 3.4 in this application* (top, right) *had a maximum of 210 bhp, good for 9.1 seconds 0-60 and 120 mph. Inside was an array of traditional Jaguar touches, including rear picnic trays* (opposite) *and classic hardwood and hides* (above, right).

very thing that had prompted the "L" sedans. One also suspects body flex was a problem; indeed, the Cs were delayed by structural rigidity from a planned 1972 introduction with the Series II sedans. A final factor, at least in the U.S., may have been image: Advertised as the "corporate sports car," the XJCs were tagged the "corpulent sports car" by more than one wag.

In any case, the coupes were phased out by early 1978, along with the short-chassis sedans. Production was always meager—1975-78 U.S. sales amounted to only 1794 sixes and a mere 686 twelves. But that only makes these Jaguars among the more collectible cars from a decade that didn't produce very many—and among the few cars that managed to look great despite Washington's best efforts to the contrary.

The next evolutionary step for the four-door XJ was the Series III, which came along in 1979 (and was bereft of the "L" suffix, the long wheelbase having effectively been standardized). This mainly involved an extremely subtle remodeling of the greenhouse by no less than the Italian house of Pininfarina. There was more rake to both windshield and backlight, a flatter roofline for still more rear-seat headroom, sharper corners in the rearmost pillars, flush door handles, bulkier bumpers—yet you had to look very closely to see the differences. A masterpiece of restrained restyling. Besides this freshened skin, the Series III XJ6 boasted a new fuel-injected 4.2 with bigger inlet valves and an additional 30 bhp.

There was also a serious attempt to solve persistent mechanical and quality troubles that, by now, had seriously tarnished the Jaguar image. The usually complimentary *Road & Track* once expressed its frustration with the Leyland-owned Jaguar of this period by calling it "a slow company specializing in fast cars."

None much faster than the XJ12, which had finally been released to an impatient public in September 1972. As discussed in the previous chapter, this model's engine was an entirely new conception at Coventry, a massive prime mover designed from the outset for displacements of up to 7.0 liters, but held to 5343 cc (326 cid) for its first appearance in a passenger car. Thanks to the use of aluminum in its block as well as its two single-cam heads, it wasn't much heavier than the old iron-block XK six. According to Jaguar, it added only about 60 pounds to front-wheel loading as installed in the XJ chassis—and that included the necessary extra cooling arrangements. What with various creature comforts and mechanical upgrades, some required by Federal decree, total weight for the XJ12 climbed above 4000 pounds. Though distribution went from about 53 percent front to about 55, road testers didn't seem to think that changed the very good XJ6 handling at all.

But the V-12 certainly did change performance. Even saddled with the notoriously sluggish, sadly standard Borg-Warner three-speed automatic transmission (optional all along for the six-cylinder cars), the muscular motor lopped two full seconds off the 0-60 time compared to the automatic XJ6; *R&T*'s figures for U.S.-spec cars were 8.6 versus 10.7. The same magazine saw top speed go from 115 to 139—and that was with the low-compression, 241-bhp "Federalized" engine. Europeans enjoyed 265 bhp and maximum velocities well into the 140s. In the somewhat lighter XJ12C, speeds as high as 148 were reported.

At an initial U.S. price of $10,800, Jaguar's new V-12 sedan cost only $1550 more than the XJ6. That looked even more sensational when you remember that this was

*Lyons acquired Daimler in 1960 and used its honorable old name on some slightly reworked Jaguar sedans. Daimler's 2.5 V-8 went into a short-tail Mark II in '62. And a long-tail six-cylinder variant was dubbed the Daimler Sovereign in '67* (this page). *A fluted grille distinguished the Daimlers from the Jaguars.*

the world's only production 12-cylinder four-door. Mercedes, BMW, even Ferrari had nothing like it. Unfortunately, the XJ12 had no sooner arrived on the American market when the first big energy crisis hit. Customers suddenly had little use for fuel "economy" of only 10.3 mpg (per *R&T*). The XJ12 was thus withdrawn from the United States after a few years and few sales, though it continued to find a small but steady audience in the UK and Continental Europe.

Uncertain fuel supplies and fluctuating gasoline prices affected those markets too, of course. Jaguar responded by reviving the 3.4-liter engine for the XJ6 and coupling it to a five-speed manual transmission from ZF in Germany. With three powertrain options—3.4 and 4.2 sixes, plus the V-12—Europeans were offered an XJ for every circumstance.

Then, an unexpected "Energy Crisis II," triggered by the mid-1979 overthrow of the Iranian shah, made even the XJ6 seem a bit thirsty. Combined with British Leyland's continuing corporate decline, that made Jaguar's future decidedly uncertain. But the British government came to the rescue by taking over the Coventry firm, thus assuring badly needed resources for further product improvements as well as a future XJ replacement. Helped by an oil shortage that soon became an oil glut, the Series III enjoyed a sales resurgence in the eighties, especially in America, long Jaguar's most important single market by far.

Genuinely improved, if still far from perfected, the Series III continued its winning ways through 1986, when the original XJ6 line reached the end of a long and remarkably successful lifespan numbering 18 years. Nearly 312,000 examples of this beautiful sedan had been

*Jaguar's long-tail compact sedan looked awkward to some and never had the sales success of its short-tail sister. Coventry tried a squared-up front with quad headlamps on its final iteration, the 420 of '67* (left), *but the results still were mixed. Under its hood was the last and the largest of the XK engines, the torquey 4.2* (top).

built—and that counts only the six-cylinder models. Twelve-cylinder figures aren't in yet. Intriguingly enough, the Series III XJ12 was still in small-scale production (about 39 a week) at this writing. Another certain future collectible? We wouldn't be at all surprised.

This Jaguar V12, as it is now called, and its badge-engineered sister, the Daimler Double Six, remain alive simply because theirs remains the only sedan platform in Coventry that can carry the big engine. A second-generation XJ6 finally debuted in 1986 (U.S. introduction: spring 1987), but the engine bay of this similar, though essentially all-new, car had been deliberately designed so that V-block engines would not fit. The alleged reason is bound up in the politics that existed during the early design stages between Jaguar people and their British Leyland masters. According to the stories, the Coventry crew feared Leyland was going to make them use an off-the-shelf Rover V-8 that they considered beneath Jaguar's dignity, so they made sure the new car could only take an inline six!

At that time, the early 1980s, it looked as though the rather "fuelish" V-12's day was over. Its popularity into the 1990s continued to surprise Jaguar product planners—and prompted them to revise the new XJ so the big engine would fit after all. The results of that work have yet to be seen as we write this, but should appear in due time.

Meanwhile, let's examine Jaguar's latest six-cylinder sedan, developed as project XJ40 but introduced with the glorious old XJ6 designation. It's very similar conceptually and even dimensionally to the well-liked original. It also looks very similar but, as one colleague put it, "with

*As the Mark VII-IX family wound down, Jaguar was readying its largest car ever, the Mark X (shown at left in prototype form). It bowed in '61 with a 3.8 six and got a 4.2 and a new name, 420G, in '64. The opulence to which buyers of the big cats had become accustomed was there from the start. In '68, its platform was borrowed for the Daimler Limousine* (opposite, top), *which remains in small-volume production today.*

all the character removed." As the first Jaguar sedan not designed by the great Sir William, XJ40 does not seem the creation of an artist but of engineers—and a committee of them at that. Its structure was simplified for easier production and made lighter in the interest of better fuel economy; systems were redesigned for improved function and reliability.

The result is undoubtedly a better car than even the Series III: a more modern car, arguably a more satisfactory car to drive and to own. And, yes, still a genuine Jaguar, though perhaps no longer a genuine sports sedan. Visually, however, something has definitely been lost. It's hard to imagine it springing from the mind of Sir William Lyons.

Compared to the old XJ6, the new one arrived on a 0.2-inch longer wheelbase, 113 inches, and was 1.2 inches longer overall at 196.4. There was a bit more interior room, though not much more, and standard equipment was upgraded in numerous ways (Bosch anti-lock braking system, a trip computer, etc.), yet overall weight was down by as much as 200 pounds. Front suspension, though new in detail, essentially repeated that of the old car. The rear suspension still relied on the lateral half-shafts to help provide wheel location but was otherwise new, with outboard-mounted brakes replacing inboard, a single coil spring per wheel (ousting the previous two-spring arrangement), and built-in anti-dive/anti-squat geometry to control pitch. The old E-type longitudinal location links were gone, massive new lower control arms taking care of that job. Also new was a "pendulum ring," a special inboard pivot joint that allowed a degree of rear-ward wheel movement to improve ride comfort over bumps without compromising wheel location, thus preserving handling precision.

Although still a six, the new car's engine was indeed all new. Only one design parameter was imposed: The block had to have the same bore centers and head-bolt pattern as the V-12, so it could take one of the V-12's heads for an economy version.

Such a single-cam, two-valve engine was offered up through 1990. With a 91-mm bore and a short stroke of 74.8, it displaced 2919 cc (178 cid). Never part of Jaguar's U.S. marketing plan, it ran on 11.2:1 compression in 1990 trim and made 165 bhp DIN (171 SAE) at 5600 rpm and 176 lbs/ft of torque at 4000 on 97-octane European leaded gas.

The primary powerplant, also called "AJ6" (the letters mean "Advanced Jaguar"), coupled that 91-mm bore with a 92-mm stroke for 3590 cc (219 cid). A classic twin-cam head maintained Jaguar tradition, but sported four valves per cylinder in late-eighties fashion. Measured against the old XK from the back of the flywheel to the front of the crankshaft pulleys, the AJ6 was 1¼ inches longer, at 34¼, and 1¼ inches lower overall, at 25½. Because of the rather massive fuel-injection system, which sat on the left side, the block was tilted to the right by 15 degrees, so the AJ6 took up more width than the more upright XK. Jaguar said that because the block and most other suitable components were cast in aluminum, the AJ6 weighed 120 pounds less than the old six, about 500 in all.

On its initial 9.6:1 compression, the new 3.6 put out 221 bhp DIN (229 SAE) at 5000 rpm and 240 lbs/ft at 4000. Early U.S. cars were considerably detuned, however,

*Forgoing the idea of marketing saloons in several sizes, Jaguar in the mid sixties began work on a single car that would cover all the bases. Lyons considered a distinctly Italianesque design* (top), *and stiffer, box-like shapes* (middle) *before arriving at his sedan masterpiece, the XJ6 of 1968* (right).

with 8.2:1 compression, catalytic converter and other pollution-control measures emasculating outputs to 181 SAE net horsepower at 4750 and 221 lbs/ft at 3750. Happily, American models were treated to a 9.6:1 squeeze for 1989, good for 195 horses at 5000 rpm and 232 lbs/ft of twist at 4000.

That still didn't silence complaints about sluggish step-off, however, so for 1990, Jaguar stroked the AJ to 102 mm, swelling swept volume to 3980 cc (243 cid). In non-catalyst European form with 9.5:1 compression (and still running on leaded gas), this new "4.0" packs 235 bhp DIN (244 SAE) at 4750 rpm and 285 lbs/ft at 3750. According to factory data, a 4.0 XJ6 with the 5-speed ZF manual gearbox available in Europe can accelerate to 60 mph in 7.1 seconds (0.3 second better than the identically equipped 3.6-liter car) and can reach 140 mph (up 4 mph).

The American version still lagged, though not as much. Initial SAE ratings for the U.S. 4.0 were 220 bhp at 4750 rpm and 278 lbs/ft at 3650. Installed in a new $43,500 Sovereign model weighing 3920 pounds at the curb (distributed 53/41 percent front/rear), it returned an average 18 mpg with standard automatic transmission. That was still the four-speed ZF automatic gearbox introduced with XJ40, and still with Jaguar's unique "J-gate" shift quadrant permitting manual selection of first, second and third gears, but newly updated with electronic shift control. As with a growing number of class competitors, this provided two driver-selected shift "programs." Press-on types use the "Sport" setting, as it delays full-throttle upshifts to higher revs than the "Normal" mode.

And what of performance? Well, *Car and Driver* reported 0-60 in 9.6 seconds, a standing quarter-mile of

*Stretched-wheelbase "L" variants of the XJ debuted in '72, including one with the 5.3-liter V-12* (top). *The original XJ6* (middle) *continued to draw acclaim for its refinement and performance. For '73, revisions inside and out created the Series II XJ. The changes extended to the Daimler-nameplate siblings.* Left: *Compare the red Series II Daimler Sovereign to the Series I car.*

17.0 seconds at 85 mph, and top speed of 133. CONSUMER GUIDE® magazine reported even better results, its 4.0 hitting 60 in just 8.9 seconds, a full second faster than a like-equipped 3.6-liter 1988 Vanden Plas model. All brisk enough for Beverly Hills probably—one sees a lot of XJ6s there—but hardly world-beating.

And that seemed to be the main trouble with the neo-XJ6: It simply did not meet an enthusiast's expectation of what a Jaguar should be. In concept, XJ40 was indeed the most modern Jaguar yet. But its development and refinement period had stretched so long—about a decade, in fact—that when finally introduced to a far more competitive world market, it seemed an also-ran. Performance was adequate by all accounts, but not thrilling by most. The same for handling and road mannerisms: satisfactory, but just not state-of-the-sedan art. The engine made nice sounds, and the cozy, leather-and-walnut cockpit was as nice a place to be as any Series III cabin. But how could a purported "sports sedan" be taken seriously if its maker wouldn't offer a manual transmission in its most important export market? The answer to that one lies somewhere between Coventry's limited resources and the seemingly limitless intricacies of satisfying the Feds.

Satisfying buyers seemed just as vexing. A prime reason for the upturn in Series III sales had been the increasing perception of the XJ6 as a lower-priced, higher-value alternative to prestige German sedans—namely those of BMW and Mercedes-Benz. Nothing new in that. Whether sedan or sports car, Jaguars have always offered superb value for money, which, of course, is the only way Sir William would have them. Both in its appointments

Right: *Britain's bobbies took a liking to Lyons' Swallow sidecars in the twenties, and thereafter, scenes like this were quite common on English byways.* Below: *the Series III sedan XJ of '79. Was it prettiest of all?*

*Coupe derivatives of the XJ sedans were offered from '75-78. Stylish inside and out, some packed V-12 punch* (top right). *Body flex and small rear seats were flaws.*

and character, the XJ40 merely continued catering to Jaguar's increasingly luxury-oriented sedan clients.

Those in America were initially offered two models: a well-equipped standard XJ6 and the plusher Vanden Plas. The latter, named for the famed British coachbuilding firm that had been folded into BL long before, was furnished with such expected "Olde English" amenities as additional leather-and-walnut trim, plus—shades of Rolls-Royce—drop-down picnic trays in the backs of the front seats. Also included were limited-slip differential, front-seat "bun warmers," even heated door mirrors. The main change for 1988 was adoption of the shorter European final-drive ratio (3.58 versus 2.88:1), another interim step to better low-end snap with the 3.6-liter engine. Despite this, fuel economy was unchanged, and the XJ6 continued with EPA ratings high enough to avoid the federal "Gas Guzzler Tax" penalty.

Offerings then expanded for 1990 to include the aforementioned Sovereign, really the previous standard model with "European-styled" square headlamps that better integrated with the squarish front than the previous round units, which were retained for a slightly detrimmed basic XJ6 priced at a very competitive $39,700. The Vanden Plas returned with little change save the 4.0-liter six, electronic transmission control, and the square lights, but was succeeded as top of the line by the limited-edition $53,000 Majestic. This was identified by a Daimler-style fluted grille (Majestic being an old Daimler model name revived), special "Regency Red" metallic paint, "diamond-polished" lacy-spoke alloy wheels, and light-tan "Magnolia" hide interior with Mulberry piping on the seats. At the same time, Jaguar answered initial XJ40 criticisms by simplifying the gimmicky trip computer, modifying the equally *de rigeur* "Vehicle Condition Monitor," and replacing the original bar-graph vacuum-fluorescent minor gauges with traditional analog dials, thus matching the round speedometer and tachometer used all along. The Majestic departed for 1991, but the three other models returned with only detail changes (and slightly higher prices).

For all this, one wishes, however idle the pursuit, that the new XJ6 had somehow embodied that old Sir William Lyons styling magic. "No Jaguar has ever been perfect," as journalists were wont to observe, but in most cases they so captivated the eye and soul that the mind forgave their little flaws and foibles. Absent the touch of the master's hand, XJ40 appeared to be a bland, cautious Anycar.

But Jaguar lovers shouldn't judge too harshly, nor lose faith. After all, it's fortunate that there is a Jaguar at all today. In the years following Sir William's departure, his company suffered some desperate times. Complete failure often loomed as a very real possibility. Yet thanks to the tenacity of many faithful old employees, and the creativity of many new ones, Jaguar managed to survive—and to do so *as* Jaguar. Most heartening of all, perhaps, are the signs that those now entrusted with its destiny really do seem to understand the significance of its traditions.

So while they labor to adapt XJ40 to accept the splendid V-12—and bravo to that—there is room to hope for even better things in the next-generation Jaguar sports sedan. Perhaps it will be a true sports sedan in the mold of the great old Mark II, the fast, handy compact that did so much for the marque on both road and track—a smaller, lighter, more nimble car that would bring out the best in the lovely AJ6 engine.

And, dare we imagine, a car so lovely to behold that we could not gaze upon it without longing? That indeed would be a genuine Jaguar.

*Concept drawings for a second-generation XJ6 were in the works from the early seventies* (far left). *By '76, Jaguar had produced a full-scale mock-up* (left) *that retained some of the early concept's taper. And by '78, a one-quarter-scale model* (below) *showed a squatter shape quite close to the finished car.*

*Introduced in 1986, the second-generation XJ6* (above) *obviously was related to the earlier cars, but its shape was less expressive. The all-new AJ6 provided power in sizes ranging from 2.9 to 4.0 liters. Inside was the usual lumber and leather, as well as a new "J-gate" shifter.* Above left: *True prowess returned with the limited-edition JaguarSport XJR 4.0.*

CHAPTER XVII

# XJ-S: FIRST-CABIN CAT

At speed, the XJ-S interior is as whisper-quiet as the first-class cabin in a British Airways Boeing. There is only a faint rushing of air over the canopy, only a mild, contented growl from the V-12. Conversation can be carried on in normal tones. The sensation of serenity is absolute. And yet the roadway is simply leaping in at you.

How fast are you going? About as fast as you care to. It might easily be 130 mph, or slower, or even faster; it's up to you and the circumstances. The car's happy at almost any speed. And so are its occupants.

That's not to say that the XJ-S is a sports car. Some people prefer to think of it as a GT, or "Grand Touring" car. In fact, it's hard to paste any conventional label on it. One of Jaguar's early marketing slogans claimed "a special kind of motoring." The XJ-S, a private jet of an automobile, fits that old description perfectly.

When British Leyland-owned Jaguar finally faced the fact that the E-type was reaching the end of the line, it also had to face a big question: What could possibly replace it? Answering that appears to have seriously troubled the firm's product planners during the late sixties and early seventies.

The original E-type was a sports car in the purest sense of the term: one derived from a race car, the Le Mans-winning D-type. As such, it was compact (to the point of being cramped), light (and thus fragile in some respects), aerodynamic-looking (but damage-prone in tight quarters), sophisticated (read: expensive to make),

*Purists recoiled when the E-type successor wasn't a hard-core sports machine. But traffic congestion had lessened the charm of the sports-car experience. And big-time racing no longer inspired road cars. So the XJ-S emerged in 1975 as a grand tourer.*

and powerful (meaning not especially economical). Yet such faults were easily forgiven as long as owners remembered that their E-types were not "Anycars" but specialized sporting implements, automobiles intended primarily to give pleasure when driven rapidly and well on open roads.

Trouble was, it was getting much harder to remember that by 1970. Denis Jenkinson's "happy carefree motoring paradise" was fast vanishing. Many of the inconveniences, discomforts, awkward details, and minor flaws of the E-type—or of Anycar—didn't seem to matter much as long as one was able to drive with exuberance. But they could swell to irksome magnitude when one was mired in slow, frustrating traffic.

Another factor was what had happened to racing in recent years. The hallowed D-type, first revealed in 1954, had represented a technical solution to the conditions, attitudes, and regulations of a day when racing sports cars were viewed as close cousins of everyday vehicles. Journalist John Bolster, for instance, thought it perfectly correct to take an actual Le Mans racer out onto an English public highway for a few hours of brisk pleasure. And, as we have seen, it had been entirely feasible for Jaguar to convert a number of D-types into XK-SSs, street-legal even in North America. The E-type of 1961 was "merely" a direct application of the company's competition knowhow to the road.

Ten years later, sports-car racing had been swept by a revolution that was as conceptual as it was technological. Cars had to be mid-engine to be competitive, but it went deeper than that. Many had come to view sports-racing cars as specialized track instruments—virtual Grand Prix racers with fenders. Only old-school diehards expected to see any design links with everyday sports cars.

Some manufacturers did try to maintain the dual-purpose "race-and-ride" tradition in the mid-engine era. Ford made a quarter-hearted effort with a roadgoing version of its first Le Mans car, the small-block GT40 of 1964, which in three different evolutions won the French enduro four years running (1966-69). Lamborghini did not tackle racing, but did proffer limited-edition two-seaters of imaginative design, stunning performance and unquestioned charisma: the V-12 Miura of 1966 and the even wilder Countach of 1973. Ferrari's 1963 250LM, which scored a Le Mans victory in 1965, was a "racer-with-a-roof" that could also be driven on the street by the truly determined enthusiast. It led indirectly to the production V-6 Dino series that began in 1967, as well as to 1976's mighty 512 Berlinetta Boxer, a street coupe capable of racing at Le Mans in highly modified form.

Within Jaguar, a few ardent enthusiasts spent a lot of time on paper evolutions of the mid-engine XJ13 prototype racer, designing several midships two-seaters for both road and track. One had a rearward-facing V-12 *à la* Countach, with the gearbox running forward to between the seats. Another, a coupe design with a more conventional rear-transaxle layout like the XJ13, went as far as a scale model for wind-tunnel development. The intent of this proposal was suggested by its simulated bumpers

positioned at the heights required in the United States, Jaguar's most important market. That created a somewhat blunt nose, and with a high-decked tail and Ferrari- and Corvette-like shrouded rear window, it lacked the loveliness of line of the Lyons era. But it doubtless would have been more practical than the E-type as an everyday vehicle of the seventies. And according to the tunnel tests, it would have been quite aerodynamically stable at very high speeds.

Would this have been the long-anticipated "F-type?" Probably, but on a sketch of it published by Robert Berry, a Jaguar factory engineer and race driver, the title was simply "Mid-Engine V12 2-Str Sports." It would have had a wheelbase of 96 inches and track measurements of 58 inches front and rear. It would have been 182 inches long, 71 wide, and a scant 41.5 high. It would have been a sensation.

Despite Jaguar's enthusiasm for another "super sports" car, British Leyland planners wanted something with far greater sales potential: a more practical, more luxurious, more commodious vehicle—a 2+2. Alas, Coventry's designers believed it was not feasible to add even small extra seats to the "Mid-Engine V12 2-Str Sports" (though Lamborghini later managed it with the Urraco, which admittedly had a much-less bulky transverse midships engine).

Another roadblock was forthcoming U.S. "barrier crash" tests. Although other manufacturers have since found that mid-engine cars can be made to pass these, Jaguar engineers thought it too difficult at the time.

Indeed, it is not hard to visualize the Jaguar staff, already demoralized by the leaden atmosphere and stormy chaos of British Leyland, just giving up their dream in disgust. The sad answer to the question, how could they replace the E-type, turned out to be: They couldn't.

Still, they did the best job they knew how to do on the so-called XJ27 program (the initials again denoting "Xperimental Jaguar") that would eventually materialize as . . . no, not "F-type" but XJ-S (with "S" allegedly for "Sport"). What they had learned in the wind tunnel from work on the streetgoing XJ13 went straight into the new 2+2's shape, notably its bluff, high-prow nose and its rather upright rear window shrouded in "sail panels" or "flying buttresses" to reduce turbulence.

However, it wasn't as easy as that, for a great deal of restless searching went on to come up with the correct visual identity for what everyone foresaw as Jaguar's new flagship. In fact, two different styling groups worked on opposing designs—hardly unusual for most automakers but a radical change for the company that for so long had been the personal fiefdom of that artistic autocrat, Sir William Lyons.

On paper, there were XJ27s with a sleek, smooth flavor similar to that of the outgoing E-type, others that evoked the brawn of certain Ferraris, even some with a

rather garish Detroit appearance. The design ultimately chosen was different yet. Frankly, it was as controversial then as it is now, and it may or may not be surprising that Sir William did have a hand in it. Though by then retired from the day-to-day affairs of the company he'd founded, he just couldn't keep away from a new Jaguar.

Lyons described his role to Paul Skilleter in the October 1975 *Thoroughbred & Classic Cars*. "We decided from the very first that aerodynamics were the main concern, and I exerted my influence in a consultative capacity with Malcolm Sayer," Lyons affirmed, referring to the company's long-time aerodynamicist, who had died in 1970 after most work on XJ-S was completed. "We originally considered a lower bonnet line, but the international regulations on crush control and lighting made us change and we started afresh."

What can be fairly said of the final shape is that, while it may not have the pure, simple beauty of some of

the great Jaguars of the past, it at least preserves their feline distinctiveness of line. And it *does* have some good angles. Moreover, you don't tire of looking at it somehow. There is never any doubt that you are looking at something different—"a special kind of motoring."

Underneath, of course, the XJ-S was entirely different from the mid-engine "super sports" in conception, purpose, and design. It was, in fact, basically a shortened XJ6/XJ12 with different bodywork, as might be divined from the model name. It even used the same inner dashboard module with its heating/ventilation/air conditioning systems, the core around which a modern car must take shape. Just as Sir William had built the XK 120 on a shortened version of the Mark V's rail-type chassis, so his successors began with an existing sedan floorpan and cut a section out of it just behind the driver's seat.

Also like the XK 120, the XJ-S wheelbase ended up at 102 inches (6.8 inches shorter than that of the origi-

*New owner British Leyland mandated that the E-type replacement be a rather practical and luxurious 2+2, so Jaguar's designers abandoned their "super sports" car dreams and worked up a series of concepts* (opposite, top). *William Lyons did influence the final car, which placed a premium on aerodynamics and meeting barrier-crash standards. Its shape has remained controversial, but also interesting. And the silken V-12 powerhouse beneath its hood guarantees that the XJ-S always will be taken seriously.*

nal XJ6). Overall length, including bumpers, was 191.7 inches, only fractionally longer than the 2+2 version of the old E-type. Track dimensions were 58 inches front, 58.6 rear, while overall body width was 70.6 inches. With a ground clearance of 5.5 inches, height came to 49.6.

Running gear, including the front-mounted V-12, was largely taken over from the four-door XJ12. The few differences: softer-rate springs to suit a lower overall weight; a stiffer front anti-roll bar (sway bar) and the addition of a rear bar to suit higher expectations for handling; and a lower numerical ratio for the rack-and-pinion steering system to make it quicker and to reduce the sedan's objectionably light steering feel.

Structurally, the XJ-S bodyshell was stiffer than the sedan's both front and rear, the better to pass those barrier tests. For the same reason, the sedan's twin fuel tanks, with one housed in each rear fender, gave way to a single reservoir positioned ahead of the trunk, well protected atop the rear suspension subframe between back seat and boot bulkhead. While developing the body, Jaguar paid very great attention to noise suppression.

The by-now familiar 5.3-liter, bowl-piston V-12 was much as before except for introduction of electronic fuel injection to the Jaguar range. This Lucas/Bosch system, a multi-port type with a squirter for each cylinder, combined with a compression ratio of 9.0:1 to give the European XJ-S 285 DIN horsepower (296 SAE) at 5750 rpm and 294 pounds/feet of torque at 3500. Many U.S.-required emissions-control measures were by now incorporated on Jaguars sold in all markets, but the American-spec XJ-S arrived with further restrictions, including a cut in compression to 8.0:1. As a result, power withered to 244 at 5250, torque to 269 at 4500 (both SAE net). Regardless of engine tune, transmissions were those well known to Series 3 E-type fans: standard four-speed all-synchro manual and optional three-speed Borg-Warner automatic.

Introduction was in September 1975, just as the world was getting back to normal after the OPEC oil embargo and the ensuing energy crisis. It hardly seemed a propitious time for any new high-performance automobile, let alone one with V-12 power, which still was quite uncommon, still the province of pricey, handbuilt Italian exoticars. Despite this—or maybe *because* of it—a new Jaguar again attracted lots of attention.

Most of it was favorable. Said the venerable *Autocar* ("The" had by now been dropped from its title):

*The XJ-S is essentially a shortened XJ6/XJ12 sedan with different bodywork. Fuel is carried in a single cell ahead of the trunk instead of in dual tanks in the rear fenders, however, and softer spring rates suit the 2+2's lower weight. Quickened steering, stiffer front sway bar, and the addition of a rear bar contribute to a sportier feel. Fuel injection was introduced to the Jaguar line on its 5.3-liter V-12, which was originally rated at 285 bhp in European tune and at 244 with U.S. emissions controls.*

"Whatever one thinks of the appearance of the XJ-S—for what personal prejudices are worth, not all of us find this Jaguar as immediately beautiful as several of its predecessors—to drive the XJ-S even for an afternoon is to admire it very much." The editors went on to describe the car as "sensationally quiet" in terms of engine, road, and wind noise. "It is certainly fast," they went on, and "perhaps best of all, it handles superbly."

Journalist Skilleter also praised the XJ-S, calling it a "unique blend of speed, comfort and silence." *Motor* termed the new Jaguar "an outrageously large and heavy two-plus-two," but at the same time "a magnificent car, not just for what it does, but for the way it does it." *Road & Track*'s Mike Knepper reported "No fuss, no bother, and the car is so quiet it is almost eerie."

Indeed, so quiet and fuss-free was this newest Jaguar that some drivers felt uncommonly removed from the driving process. "The car is quick," said Patrick Bedard in *Car and Driver*, "but it denies you the thrill of speed. It never seems to go fast, even when the speedometer needle is pointing out three-digit numbers. It's too composed to seem fast." He went on to call the XJ-S "extravagant" because, at least in an America newly yoked by a national 55-mph speed limit, the car had "one outstanding quality—poise at extraordinary speeds—and you can't use it."

About those speeds. *Autosport*'s John Bolster tested a 285-bhp manual-transmission car weighing "1 ton 14 cwt," or about 3800 pounds, and equipped with a 3.07:1 rear axle. His results showed 0-60 mph in 6.7 seconds, quarter-mile performance of about 95 mph after approximately 15 seconds (the data were presented only in graphic form), and a top speed of 154 mph at 6250 rpm, a comfortable 250 rpm below redline. Fuel consumption ranged between 10.8 and 15 miles per U.S. gallon.

*C/D*'s "Federalized" 244-bhp automatic model was listed with a curb weight of 4020 pounds (distributed 56.3/43.7 percent) and a 3.31:1 final-drive ratio. Its 0-60 time was 7.5 seconds, its quarter-mile dash 16.0 at 90.3 mph, and its observed maximum speed 140. Gas mileage was 12.0 mpg U.S., and list price in January 1976 was $19,200, which looked like a remarkable value against the specification. Another Jaguar tradition maintained.

As to its character on the road, the XJ-S pleased most of those who drove it without actually kindling their fervor. It seems to have been regarded more as an admirable, if "extravagant," traveling appliance than as an exciting traveling companion. "I like the XJ-S," Knepper told *R&T* readers, "but my enthusiasm doesn't go much beyond that." In his opinion, Jaguar had refined too much personality out of the car; the XJ-S failed to "set the motoring world on its ear" as had the E-type and XK 120.

*C/D*'s Bedard was more upbeat. "The suspension is amazing," he declared, "taut but not hard, and the geometry is accurate. The car seems immune to the laws of physics. It makes you think you could never lose it—that it would compensate automatically for your blunders." That feeling came from a predictable, progressive understeer in corners that was only compromised by a modest change in attitude if the throttle were lifted—"Not much, but enough so that you know it must have a driver." Bedard summed up the handling by saying, "The car has no tricks. It is merely your obedient servant, with capabilities that should be talked about rather than used."

Bolster's assessment for *Autosport* went like this: "Near-neutral handling and exceptional traction are its outstanding characteristics. It is ultimately possible to make the rear end break away on a test circuit, but only by driving much harder than would be in any way reasonable on the road. The behaviour continues to be safe and predictable on wet surfaces and the steering is very much better than the rather over-assisted setup on other Jaguars, though there is still not much feel of the road."

It wasn't for its handling that one would buy this car, Bolster seemed to be saying, but for its effortless straight-line capabilities. "The whole point of owning an

XJ-S is that it is entirely unstressed at any normal speed, and stability which has been worked out for 150 mph is incomparably better than that of ordinary cars. It is utterly safe, and in an emergency it can be braked to a standstill in an incredibly short distance . . ."

But: "These speeds are somewhat academic nowadays and in case somebody accuses me of advocating such driving, perhaps I should utter a word of warning. Anybody who is reasonably competent should be able to drive safely at speeds up to 130 mph, but though much higher velocities are not at all dangerous in themselves, they call for judgement which can only be acquired by practice. The man who seldom drives at much more than 100 mph should think twice about going up to 150 mph without some exploration of intermediate speeds. If things go wrong, it all happens a bit sharpish, so don't say that uncle Bolster didn't warn you!"

And if the XJ-S wasn't really a sports car, cut Jaguar some slack, said the veteran *Autosport* writer. "From time immemorial," said Bolster, "the enthusiast has bemoaned the tendency of manufacturers to turn away from 'hairy' sports cars toward ultra-luxurious GT coupes. The answer is that everybody longs to own a stark semi-racer, but nobody actually buys them."

No, people who could afford to stock their garages with really fast road machinery were buying big, heavy, costly Aston-Martins, Ferraris, Lamborghinis, Maseratis, and Mercedes-Benzes. Offering a car of the same nature simply looked like good business to Jaguar—which clearly remembered having to discount the last of its enthusiast-oriented E-types to get rid of them.

Alas, part of good business is building a quality product. And as part of the flailing—and failing—British Leyland conglomerate, Jaguar had trouble with that in the late seventies. Bodies were one very sore point, arriving from an outside supplier with all manner of structural faults. While this situation was eventually improved, it contributed to a sales picture that forced Jaguar to suspend XJ-S production for a few months in 1980. For a time, in fact, the car's future was in serious doubt. The onset of a second fuel crunch the previous summer hardly helped.

Which brings us to a significant factor in saving the XJ-S: a completely new cylinder head that improved the V-12's fuel mileage by some 20 percent. The patented work of a gifted and versatile Swiss engineer named Michael May, this so-called "Fireball" design retained the single-cam, parallel-valve configuration, but moved the combustion chamber out of the piston and into the head. The valves were on two different levels, the inlet lower than the exhaust, and the resultant asymmetric chamber was carefully shaped to promote turbulence as an aid to rapid and complete combustion of a lean air/fuel mixture.

Jaguar had already revised the fuel injection and raised the older head's compression ratio to 10.0:1. Now, with the Fireball head and a very powerful ignition system, compression went to a remarkably high 12.5:1 for European engines. That put horsepower up to 299 DIN (310 SAE) at 5500, and torque to 318 lbs/ft at 3000. Even for low-octane America, compression was boosted to a stout 11.5:1, which gave SAE horsepower of 262 at 5000 and torque of 290 at 3000.

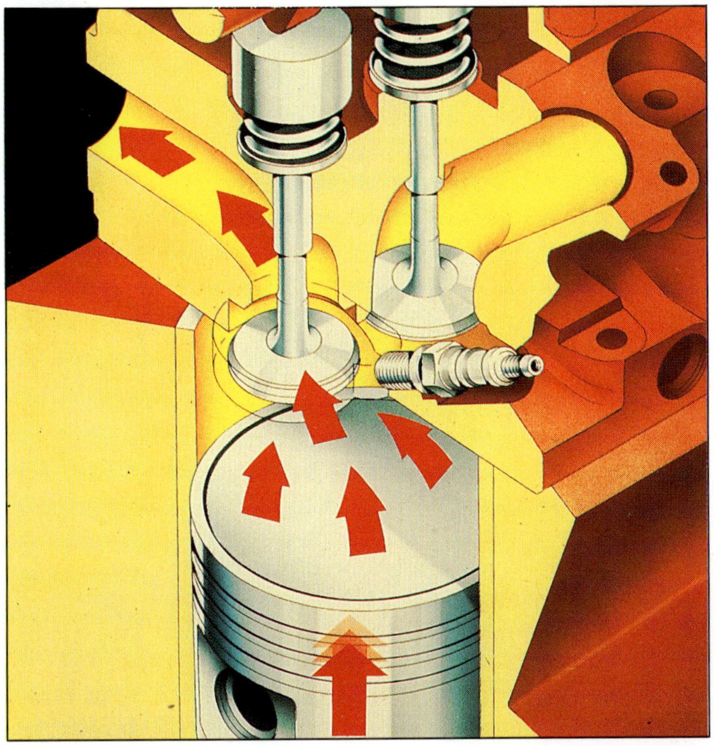

*Flying buttress roof is a signature, while European-market headlamps look more pleasing than the quad-sealed-beams on American cars. Prominent hood bulge on the coupe pictured identifies it as among the non-U.S. XJ-Ss with the twin-cam 24-valve 3.6-liter six from the second-generation XJ6 sedan. The illustration at left details the fuel-saving cylinder head introduced in '81.*

Introduced in 1981, this updated engine was accompanied by a revamping of the car itself. Walnut veneer and real leather graced an interior that was originally vinylized. Outside were cast-alloy wheels of a new design and other trim differences, including chrome-plated bumpers and a painted "coachline" accentuating the length of the car. The stronger engine allowed use of a taller, more fuel-efficient 2.88:1 rear axle with no loss of performance. For the next few years, the whole package was identified by large chrome letters on the tail reading "H.E."—"High Efficiency."

Curiously, though, the only transmission available now was the General Motors THM-400, the smooth-but-sluggish three-speed Turbo-Hydramatic. The facts were that only a tiny proportion of XJ-S customers had chosen the old manual, and these do-it-yourself diehards unfortunately had discovered that it had been overstressed and not very reliable in this application. Though Jaguar's Harry Mundy went to the trouble of designing a five-speed strong enough for the V-12, his bosses decided to forget about it.

Returning to open-air motoring was higher on their priority list. Although a convertible had been part of the original XJ27 plan, subsequent moves in American safety circles generated a widespread anticipation that ragtops were going to be outlawed in that vital sales arena. Though this never happened, it took Jaguar a long time to come back to a drophead XJ-S.

It finally was introduced to Europe and Britain in October 1983. Called Cabriolet but badged "XJ-SC," it was actually a semi-convertible. It had fixed side-window frames and a sort of perimeter roof cum Targa bar that remained when two rigid top panels were removed (they stored in the trunk) and a flexible back "hood" was folded down. It looked a bit ungainly opened up, but building a drop-top this way enabled Jaguar to preserve most of the coupe's structural rigidity without heavy extra bracing. No "plus-two" seats were fitted, by the way; the Cabriolet was strictly a two-seater.

Mechanically, the XJ-SC was interesting in that it was first offered with only one powerplant: the new all-aluminum, twin-cam, 24-valve AJ6 "slant six" destined for the upcoming second-generation XJ6 sedan (née XJ40). That 3.6-liter, 225-bhp (DIN) engine was also offered as an option on the XJ-S coupe, though not in the U.S. In both "XJ-S6"s the transmission was manual, a modern five-

*Jaguar withheld a drophead XJ-S until 1983, when it released the two-seat Cabriolet* (opposite). *It had fixed window frames, a targa bar, two removable rigid roof panels that stored in the trunk, and a folding rear-top section. True open-air motoring didn't come to the factory-production XJ-S until '88 with introduction of the Convertible* (above). *Absence of the Cabrio's fussy framing required heavy, performance-robbing chassis bracing, but no XJ-S is more beautiful.*

speed supplied by the German ZF company. A distinguishing visual feature of these cars was a somewhat more prominent "power bulge" in the hood, needed to clear the slightly taller engine.

Though the six-cylinder XJ-S was not a true performance car, it is interesting to look at its abilities. In coupe form it weighed about 3640 pounds not quite empty of fuel, or about 4050 as tested by *Autocar*, meaning it was around 200 pounds lighter than the companion V-12 model. Proportionately, that was less of a drop than the reduction in engine capability, so performance suffered by comparison. The 0-60 time was 7.2 seconds, the quarter-mile was covered in 15.6 at 92 mph, and top speed came to 138. Certainly the baby XJ-S was no slouch, but neither was it quite as sparkling as its big sister.

Nor was the AJ6 engine as well-received as Jaguar had hoped. Press reports labeled it "lusty," but some also called it "harsh" and rough-running. Of course, any engine would seem coarse compared to the glass-smooth V-12, but in fact there was a design problem: the aluminum crankcase wasn't strong enough to keep the long, heavy crankshaft firmly in its place. As it whirled around, it wobbled in its bearings and set up vibrations the occupants could feel. Jaguar went back to work, and came up with fixes in time for XJ40's introduction, so the idea of trying out this engine in limited production with the XJ-S definitely paid off.

Meanwhile, Jaguar felt it necessary to start offering the V-12 as an option in the Cabriolet, starting in 1985. Again, though, this XJ-SC 5.3 came only with automatic transmission. It was the only Cabriolet that would make it to the States, sales commencing in spring 1986. It wouldn't last long, but Jaguar did go to the trouble of offering an optional rigid rear roof section with heated glass backlight that could be snapped on in lieu of the cloth top.

The reason the Cabriolet didn't last long in America was the true full convertible that effectively replaced it in the spring of 1988. Jaguar knew the demand was there, because for a year it had been cooperating with an outside "chop shop," Hess & Eisenhardt in Cincinnati, which converted standard coupes into full dropheads under contract. Jaguar thought enough of these to back them with its full warranty. Now, with design assistance on the folding top from Karmann of Germany, Jaguar took over the program itself.

Top down or top up, the Convertible certainly was the best looking XJ-S to date. Shorn of its awkward "flying-buttress" cabin, the lower body's long-limbed grace shone forth. On the downside, the added belly-bracing required for chassis strength inevitably increased overall weight—by about 220 pounds—so performance was off again. But you can't have everything, and for most people a couple of hours of sunny top-down cruising can very nearly compensate for a couple of tenths of a second lost in acceleration.

Besides, the hardest core of the speed-seeking fraternity was now being served, and served well, by a surprising number of aftermarket entrepreneurs who had seized upon the XJ-S as the basis for some truly stunning high-performance road machines. One was the German tuning firm Koenig, which turned it into a monstrous-looking supercar. Of the other "modifieds," probably the best-known were the products of Britain's Brian Lister, who back in the fifties had used Jaguar engines to power the series of very fast, and very successful racing sports cars bearing his name.

Thirty years later, "Lister-Jaguar" denoted an XJ-S coupe which in some cases had V-12s taken out to the full design potential of seven liters. With associated changes to tires, wheels and suspension components, plus appropriate aerodynamic aids and sometimes even the addition of a very expensive, but very strong, Getrag five-speed manual, the result could be a veritable supercar. By 1989, the Lister had a bore and stroke of 94×84 mm for 6996 cc (427 cubic inches), and specially made big-valve cylinder heads that reverted back toward the old pre-Fireball design. Horsepower was a massive 496 at 6200 rpm, torque a stupefying 500 lbs/ft at 3850. According to the British magazine *Fast Lane*, 0-60 time was 5.1 seconds on an effective final-drive ratio of 2.33:1, and top speed was projected—perhaps fortunately, it was not actually observed—at something on the scary side of 200 mph.

Once again, Coventry detected a trend and decided it had better offer a higher-performance XJ-S itself. But not all by itself, in this case, for it called on the expert hands who worked for one Tom Walkinshaw.

A burly, intensely competitive Scot, Walkinshaw had been a racing driver in the early eighties when he began to gain the factory's ear with bold, but sensible, plans to campaign the XJ-S in the European Touring Car Championship.

*Though to the manor born, the XJ-S could play rough. Group 44's 195-mph example captured the 1978 Trans Am championship in America, and Tom Walkinshaw Racing took the '84 European Touring Car title with one* (below). *The TWR effort inspired a joint venture with Jaguar to create the production-based XJR-S* (opposite). *Its 318-bhp 6.0-liter V-12 was good for 6.5 seconds 0-60 and a near-160-mph top speed.*

Jaguar knew the XJ-S had racing potential because of its American performances in the late seventies. Bob Tullius' Group 44, which had achieved so much success with the V-12 E-type in SCCA B-Production racing in 1975-76, had then turned to the XJ-S for the same club's revitalized Trans-American series. In Group 44's hands, the ultra-refined road coupe turned into a raging racer with a 560-bhp engine that could push it to a genuine 195 mph—and that did push it to the Trans-Am championship in 1978.

Taking up the European Touring Car challenge partway through the 1982 season, Walkinshaw's race-prepared Jaguar was tuned for "only" 400 bhp, but immediately threatened the previously all-conquering BMW 635 CSi racers. After another two seasons of hot competition, the now-close partnership of Jaguar and Tom Walkinshaw Racing (TWR) wound up 1984 with the series title. Walkinshaw capitalized on the success with a hot-rodded roadgoing XJ-S that he sold in very small numbers as the "Jaguar Sport." Meanwhile, he moved his racing effort up to Group C, using the potent Jaguar V-12 as the basis for a series of mid-engine two-seaters aimed at Le Mans. The program achieved a magnificent victory in the 24 Hours in 1988—Jaguar's sixth at the Sarthe circuit—and again in 1990.

All this led the factory to join forces with Walkinshaw in the spring of 1988, when JaguarSport was formed as a new joint-venture operation. One of its chief duties was to create low-volume production-based models such as a special-edition XJ-S much like Walkinshaw's original "Sport" with a 6.0-liter version of the V-12. Accordingly, certain coupes and convertibles were earmarked on the Browns Lane production line for special treatment—big wheels, stiff springs, and that new long-stroke engine—before being shipped for detail completion at TWR in Kidlington, a town south of Coventry near Oxford.

The result was an official factory-sanctioned scorcher called XJR-S. (A similar treatment for the neo-XJ6 sedan produced an "XJR 4.0 Litre.") In 1990 guise it wore a special combined front bumper cover and airdam, rear bumper with lower valence, a decklid spoiler, aerodynamic rocker-sill extensions, unique badging, and big ZR-rated tires—225/50s in the front and 245/55s in the rear, mounted on special 8 × 16 aluminum wheels. There were also interior and suspension differences, of course, and black satin-finish exterior trim was available on special order to replace the normal chrome. An automatic remained the only transmission, though its shift points were recalibrated to suit the enlarged V-12.

That 6.0-liter engine was naturally the R's primary selling point. A new crankshaft gave an increase in stroke from the normal 70 mm to 78.5, thus raising displacement from 5343.8 cc to 5992.7 (365.6 cubic inches). Accompanied by various other changes, including an electronic engine-management system borrowed directly from contemporary racing practice, the 12-percent increase in capacity gave a 17-percent boost in torque over the 1990 standard engine, from 310 lbs/ft (at 2800 rpm) to 363 (at 3750). The new horsepower rating was up 11 percent, from the 5.3-liter 1990 car's 286 (at 5150) to 318 (at 5250).

According to the factory, acceleration to 60 mph could now be achieved in 6.5 seconds, and top speed was "almost" 160 mph. At this writing, Jaguar was working on certifying the 6.0-liter for U.S. sale, but the work was going slowly and the engine has yet to appear here.

While listing XJ-S variants, we should mention one of the least likely, but arguably most successful, private conversions. This is an extended-roof "sport wagon" from a small British enterprise named Lynx, well-known for its faithful D-type replicas based on E-type components. Named "Eventer" (in the sense of attending events, such as riding to hounds, don't-you-know), this not only made the closed XJ-S into "the world's fastest Estate Car," as Lynx claimed, but certainly improved on the original looks.

Perhaps Jaguar had taken notice of this trend as well. Not that Browns Lane was preparing to mount a challenge in the station-wagon field, but it was widely understood by 1990 that a modestly facelifted XJ-S was on the way as a 1992 model.

However, neither did the company feel any need of unseemly haste. As we found on a drive through England in a standard 1990 XJ-S V12 Coupe (to use its official designation), Jaguar's grand-touring 2+2 remains popular in its native land 15 long, eventful years after its introduction. Astonishingly popular, in fact. It is still a literal headturner, a car that sometimes actually halts strangers in their tracks and that often brings cordial acknowledgment from other XJ-S drivers. On two occasions, its air of opulence and breeding induced the owners of small country hostelries to offer us choice—and prominent—parking spaces by their main entrances. "It'll do the hotel good!" one exclaimed.

One feels so well-handled in the finer establishments of Britain, and our journey in the Jaguar was in that splendid tradition. The car acted as a master host, entertaining its honored guests in comfort, grace, and dignity.

It still is no sports car despite the many changes it has been accorded over those 15 years. As ever, the wonderful V-12 churns out power like a hydroelectric plant, but it's simply too quiet and unobtrusive to generate excitement. An unresponsive automatic gearbox, numb power steering, and vast physical bulk conspire to rob the car of any willingness to romp through the narrow, curving lanes that have bred great British road cars for generations. Great weight all too readily overwhelms the standard tires, thus further dampening your sporting inclinations. And the interior is replete with ergonomic details that seemed old-fashioned, or just plain awkward in the seventies, never mind the nineties.

With patience, though, and given the discovery of the right circumstances . . . Well, perhaps *The Autocar*'s assessment of the SS Jaguar 100, now half-a-century old, is just as apt for the Jaguar XJ-S: "An outstanding feature recalled . . . was that, whilst one felt sufficiently at home in it from the beginning, there came a stage, after perhaps a couple of hundred miles, where one suddenly found a great deal more in the car than there had seemed to be at first—not so much in sheer performance as in confidence in it." As with the grand old SS 100, those who take time to get to know the XJ-S usually find that it grows on them.

Where the modern cat proves outstanding is out on the open Motorways, where it eagerly puts its shoulder to the wind and devours miles with a relentless appetite. The firm leather seating, the elegant polished walnut, the overall mechanical and aerodynamic refinement all add up to a marvelous sense of serenity and security. It is indeed the same commanding feeling of absolute luxury one experiences in a private jet.

In its element, the Jaguar XJ-S was, and is, truly first-cabin travel. What a way to go!

*High-performance variations on the XJ-S theme: Britain's Lister-Jaguar* (opposite, left) *and the wild Koenig treatment out of Germany* (opposite, right). *The more typical XJ-S exudes prestige and coddles with fine woods and leathers. And while it's a heavy car and dated in some respects, few machines feel more capable over long distances at great speed.*

CHAPTER XVIII

# Revolution in Racing: Super Sports Jaguars Enter the Modern Age

Many people, even some who call themselves automotive enthusiasts, see racing as a trivial pursuit. But the hard core of genuine car people at Jaguar have always believed that popular enthusiasm for their marque arises from its performance, and that the pinnacle of performance is achieved through racing.

Never mind the other good qualities built into Jaguar cars. The company could never have reached its standing with its ardently supportive public—and thus in all likelihood would not survive today—had it not been for the positive performance image conjured by a serious-minded commitment to top-line international sports-car racing.

The payoff, of course, was the spectacular quintet of overall victories in the world's most prestigious sports-car event: the 24-hour endurance contest at Le Mans. As

"Lofty" England, then Jaguar's racing team manager and later its managing director, once said, that legendary success in France during the fifties "put us on the map." The fact that the all-conquering C- and D-type racers won with only slightly modified versions of Jaguar's everyday six-cylinder XK engine was not lost on the buying public. This perception had a very real and vitally important effect on the way Jaguar owners viewed both their machines and themselves.

Yes, much has changed since then. Road cars and racers, once close cousins, have followed such diverse evolutionary paths that it is now often very difficult to see any relationship at all. Yet Jaguar's most recent triumphs at Le Mans—its joyous sixth and seventh victories in 1988 and 1990, masterminded by Tom Walkinshaw Racing—were achieved with highly developed *but still recognizable* versions of the regular production V-12.

Good enough, but now something has happened that once again vests Jaguar's racing successes with relevance for today's highway drivers. In 1990, the company's high-performance subsidiary, Walkinshaw's JaguarSport, announced plans to create two new mid-engine "super sports" two-seaters for sale to the enthusiast public.

One, labeled XJ220, was originally shown as a prototype with Jaguar's venerable V-12 block wearing special dual-cam, four-valve heads. In production form the car will be shorter and lighter than the prototype and will use a detuned version of the modern turbocharged V-6 racing powerplant that Jaguar has recently campaigned in short-distance events both in the United States and Europe.

That will be strictly a street machine, but Jaguar's other new sports car is a thrillingly dual-purpose vehicle in the wonderful old "race and ride" tradition. Called R9R in prototype form and renamed XJR-15, it is based directly on the Le Mans-winning XJRs—an ultra-high performance "spec" racer that is also street-legal. Well, some streets, anyway. Sadly, it is never likely to be welcomed on American ones. More than adequately powered by a 6.0-liter version of the single-cam-per-bank V-12, the very-limited-production XJR-15 carries a list price of close to one million U.S. dollars.

Both these mid-engine developments are important not only in themselves but for what they imply—namely, that racing as a wellspring of sports cars has not gone dry. As both will incorporate much of the performance expertise accumulated by Jaguar and TWR engineers during their past half-decade of cooperation, it is fitting to review the enormous evolutionary strides made by racing cats since the distant days of the D-type.

When Jaguar withdrew from direct participation in international competition after Le Mans 1956, it had no intention of allowing its motorsports interests to plunge into a dark age. Yet that's precisely what happened. Behind the scenes, some company engineers determinedly nursed the racing flame with various ventures: the E2A variant of the E-type concept, originally intended to succeed the D-type at Le Mans; the Lightweight E-types, which did race, though too late to be very effective; the mid-engine XJ13 prototype and several similar exercises, all of which long remained secret. But for one reason or another—the 1957 factory fire, later economic problems, Jaguar's eventual sale into conglomerate slavery—the competition flame was left to smolder as a mere ember for nearly 30 years.

Though it never raced, the XJ13 is the key to the modern mid-engine story. Very ambitious, it began on paper as early as 1955, when, according to Paul Skilleter's research, engineers drew up their first serious plans for a V-12. As with most Jaguar powerplant projects, the ultimate duty station of this new prime mover would be under the hoods of sedans. But William Heynes and his staff had already learned a great deal about their XK engine by adapting it to the rigors of racing. Think how much would be learned, they must have postulated, by starting with a racing engine, then adapting it for highway service. If it proved strong and durable enough to win Le Mans, or so the reasoning continued, it should be both brilliant and bulletproof on the street.

Sound thinking, but not until late in 1964—eight years after the company went on its "temporary" racing sabbatical—was the first V-12 actually built and running on the dyno. Claude Baily was chief project engineer. To meet appropriate racing regulations, he specified displacement of just under 5.0 liters: 4993.5 cc (304.7 cubic inches) from an 87-millimeter bore and a 70-mm stroke ($3.425 \times 2.756$ inches).

With two banks of six cylinders arrayed at 60 degrees, the block/crankcase was cast in aluminum alloy. So were the heads, of course. These closely resembled the familiar twin-cam XK design, but with two main differences. One was the included angle between the valves: also 60 degrees, and thus noticeably narrower than the 75 of the "35/40" D-type head. The other was the way the intake ports came down from the top of the head between the valves; it created a magnificently bristling visual effect as well as theoretically superior breathing. Fuel injection was featured, as was dry-sump lubrication. On compression of 10.4:1, the Baily V-12 showed 502 horsepower at 7600 rpm and 386 pounds/feet of torque at 6300. Some reports suggest further development realized about 10 percent more power. Jaguar stated dry weight of the

*Coventry's blood always ran hot for racing. Early success in competition—crowned by five '50s Le Mans wins—swelled Jaguar's public image. But the gap between track and road machines widened in the '60s and '70s, and it was the early '80s before the factory rejoined the fray—and '84 before it returned to the French classic.* Left: *an XJR-6LM in the pits at Le Mans, 1986.*

finished quad-cam as 648 pounds.

Work on a car to carry this engine didn't start until mid 1965 and wasn't finished until early the following year. Recorded testing didn't begin until March 1967. Obviously, the builders had put their souls into the new racer, but their company had no heart to race it. The main problems were probably want of time, personnel and money. As has often been said of the Lightweight E-types, the XJ13 might have had a chance had it gone on a track at the earliest possible moment. By the time it did run, it was several seasons obsolete.

Still, it was an interesting exercise—and remarkably advanced considering the neglect it suffered from top management. It was certainly modern. The V-12 sat just behind the two-seat cockpit and, despite its considerable length, it was mounted longitudinally, a layout that took up so much wheelbase that the cockpit turned out quite cramped. But XJ13 was at the forefront of contemporary racing science in incorporating the engine block as a stressed chassis element. Also per mid-engine custom, the engine drove to a five-speed rear transaxle, here supplied by ZF.

The rest drew heavily on Jaguar's experience in both construction techniques and aerodynamics. Very familiar was XJ13's sheet-aluminum monocoque clothed in another of Malcolm Sayer's masterful, smoothly rounded, hand-hammered aluminum bodies. Suspension was all-independent, of course. The front actually used some E-type hardware, although coils substituted for torsion bars as the springing medium. At the rear, halfshafts formed part of the wheel location linkage, as on the E-type, but XJ13 again followed contemporary racing practice in having a pair of long trailing links each side for more positive geometric control. There was only one coil spring/shock absorber per rear wheel, instead of the E-type's two. Nor was there anything like the E-type's separate rear suspension subframe, and the rear disc brakes mounted outboard instead of next to the differential. All four discs were ventilated—and massively sized at 12×1.25 inches. Wheels were 15 inches in diameter with an eight-inch width front, 11-inch rear. The radiator was forward, D-type style, but the rubber fuel cells, holding a total 49.1 U.S. gallons, were tucked into structural boxes either side of the cockpit instead of in the tail. Though open, that cockpit was snugly wrapped by a full-width, metal-framed windshield and high side windows.

Differing dimensions have been published over the years, but a thorough 1973 *Autocar* report states that the XJ13 was 176.5 inches long and 73 wide. With its four-inch ground clearance, overall height was only 38.5 inches at the top of the windshield frame. Wheelbase was 96 inches, track 56 front and rear. Jaguar gave out empty weight as 2180 pounds (the engine/transaxle accounted for 886), distributed 41 percent front, 59 rear. "Kerb" weight came to 2478 pounds.

XJ13's entire purpose was closed-circuit road racing, where conventional standing-start acceleration is irrelevant, so apparently no such numbers were ever taken. We do know it ran a record average lap of 161.6 mph around the banked test track at England's Motor Industry Research Association (MIRA), but could only reach about 175 on the short straightaways there. To keep

pace at Le Mans with the Ford GT40s and Ferrari P4s of its day, it would have had to top 200 down Le Mans' ultra-long Mulsanne Straight.

The sleek, powerful XJ13 might well have been capable of that. However, Jaguar would have had to put a lot more money and effort into a racing program for the late sixties than had ever been required in the fifties. And in all likelihood, Sayer would have had to compromise his lovely, aerodynamically "pure" body shape with spoilers and trim tabs.

As things turned out, the shape was compromised by a crash. In January 1971, after nearly four safe years under a dust sheet in a corner of the factory experimental department, the gracefully aging racer was reawakened to appear at MIRA in a sales film centered on the upcoming Series 3 V-12 E-type. The movie-making went well enough, but some extemporaneous, just-for-the-hell-of-it laps later on were too much for one of the tired old mag racing wheels, which disintegrated at 140 mph, sending the sole XJ13 tumbling violently into near-destruction.

Luckily, nobody had gotten around to scrapping the original wooden metal-forming bucks, so Jaguar was able to build a new bodyshell. Restored to running condition, XJ13 was finally unveiled in a public demonstration at the British Grand Prix at Silverstone in 1973. The beautiful thing survives today, an honored resident of the small museum at the Browns Lane factory and an undoubted focus of countless reveries about what might have been.

For what did happen, the scene shifts to America. From its headquarters in Winchester, Virginia, the racing enterprise Group 44 had long been doing superb image enhancement for the British Motor Corporation (BMC) by winning numerous races and championships with various Triumph sports cars. After Jaguar became part of the conglomerate under the British Leyland label, Group 44 turned to the V-12 E-type. Driving a highly race-prepared open two-seater, company head Bob Tullius vanquished many small-block Corvettes during 1974, and wound up the 1975 season as national driving champion in the Sports Car Club of America's B-Production class. Turning then to SCCA's Trans-Am "sedan" series with the new XJ-S, Tullius and his team accumulated a similar record. By 1982, Group 44 and a revitalized, newly independent Jaguar were cooperating on an all-new sports-racing car for an assault on Le Mans.

The attack was actually two-pronged. Le Mans cars then had to be built to European Group C specifications written by the FIA (International Automobile Federation). These were quite similar to regulations governing America's GTP or "Grand Touring Prototype" class administered by the International Motor Sports Association (IMSA). Group 44 therefore designed a GTP car that would be developed in IMSA competition but taken to Le Mans—a long-held personal goal of Bob Tullius.

After so many years of race-winning experience with the single-cam-per-bank production V-12, Group 44 had a lot of confidence in it. Creating a chassis to carry it was the job of respected American designer Lee Dykstra. The first car was ready for testing in June of 1982. Designated XJR-5 as Group 44's fifth distinct racing model,

*Jaguar developed the mid-engine XJ13* (below) *in 1965 to carry its new racing four-cam V-12* (opposite). *The engine displaced just under 5.0 liters and made 502 bhp at 7600 rpm.*

*The XJ13's cockpit certainly looked race-ready, though the only action it saw was in testing. The first XJ13 was destroyed in a 1971 demonstration run, to be rebuilt as, literally, a dream car.*

counting the E-type racer as the first and various generations of Trans-Am XJ-S as the next three, it was mid-engined, of course, and at first glance exhibited certain other similarities to XJ13: a chassis containing much aluminum, engine block as a stressed chassis element, bodywork unusually good-looking in an era of comparatively graceless wedges.

In detail, however, Dykstra's design had nothing in common with Coventry's old exercise. Much of the aluminum chassis was crafted of ultra-light but super-strong bonded honeycomb. There was also a good deal of that space-age composite material, Kevlar. The whole structure had been envisaged from the start not as a roadster, but as a coupe with a very strong integral roll-cage. Furthermore, chassis and body shapes, bottom as well as top, were designed as an aerodynamic team to generate a powerful downward force to enhance high-speed tire adhesion.

The last point highlights the revolutionary changes that had so widened the technical chasm between racing and road cars. In the sixties, some chassis designers began placing engines behind their racing cockpits, correctly reasoning that extra weight there would improve traction for both acceleration and braking. It did. But they also found that this "midships" configuration made for more responsive handling and faster cornering. That encouraged tire engineers to make wider, stickier treads that could take better advantage of the new chassis capabilities.

At the same time, aerodynamicists began drawing spoilers, wings, "racer's wedge" bodyshells and finally ground-effects underbodies to put extraordinary dynamic loads on the much-more-capable tires, thus further increasing grip. This, in turn, prompted a general lowering of ride heights to only fractions of an inch above the road, which required springs so stiff that the cars rode like giant go-karts. One result of all this was that some classic racing circuits were abandoned while others were reconstructed to suit the new cars.

Within a bewilderingly short time, therefore, the "racing sports car" changed radically. Instead of an ultra-high-performance limited-production automobile competing on closed loops of normal roads, it became an ultra-specialized competition instrument, optimized for slashing around flat, smooth, increasingly artificial tracks with generally short straights and tight corners. Driving such a device out the gate and onto the open highway, as in the old days, would be the silliest of stunts.

*Road & Track* proved the point with a September 1983 track test of an early Group 44 XJR-5. The car was not only trucked to the Road Atlanta venue but, more tellingly, had a five-speed transmission still geared for the much longer straights of Daytona (77 mph in first), so it couldn't even struggle creditably through the standard acceleration runs. Another problem, as tester Joe Rusz recently recalled, was an unshielded ignition system that electronically interfered with the magazine's computerized test equipment. Rusz could thus "only speculate" that, based on prior tests of other vehicles, a properly set-up GTP Jaguar would do 0-60 mph in under 3.8 seconds and the quarter-mile in something better than 11.5 seconds at 131 mph. True top speed was also unmeasurable,

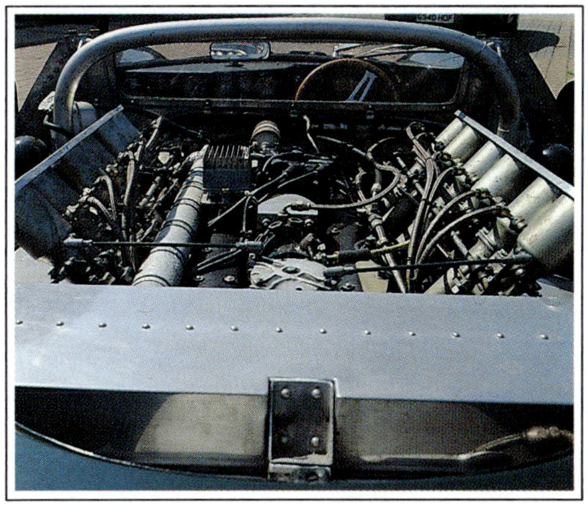

*Mounted longitudinally behind the two-seat cabin, the fuel-injected all-aluminum V-12 was a stressed chassis element. It pushed the 2478-pound XJ13 to 175 mph, but the car would have had to top 200 to keep pace with the Ford GT40s and Ferrari P4s of the day. Cash-strapped Jaguar would have had to pour a lot of money into a race program to make it competitive.*

but would have been 219 mph had the car been geared to reach 7500 rpm in fifth.

Driving impressions were similarly sketchy, partly due to cold weather and partly because the cockpit was best suited for taller drivers—a first for a racing Jaguar! Still, Rusz was able to work up to a "seven-tenths" pace: "At slow speeds the steering is heavy and quick and it takes some doing to keep from weaving down the track. But as I adjust to the car and pick up speed, the steering improves, getting lighter and precise. I go faster and realize that the car sticks to the pavement, partially because of its ground-effects design, which is just beginning to work, but mostly because of its suspension and those very wide Goodyear tires. On occasion, I hesitate going into a turn and the car feels loose. But a blast of power plants its tires firmly on the roadway and sends the car scurrying off to the next turn." Rusz still vividly recalls the Jaguar's "razor-sharp" engine response and its surprising nimbleness. "It drove not at all like a big car, which it was."

How big? History is a slippery business, and some of the magazine's published specifications came from inaccurate press materials. Happily, Group 44's private records, made available to us by long-time technical chief Lawton "Lanky" Fouchee, confirm that all the cars had wheelbases not of 102 inches as erroneously reported, but 108.5 inches. That was several inches longer than that of any other contemporary GTP—in fact, it was only three-tenths of an inch shorter than a first-series XJ sedan.

Basic layout was conventional for the time: a mid-engine coupe with two seats jammed together on the chassis centerline, the driver on the right and the left seat seldom, if ever, occupied. The fuel tank was sheltered safely just behind the cockpit. The long V-12, again mounted lengthwise and braced by steel tubes, actually formed the rear third or so of the chassis structure. Drive to the rear wheels went through a British-made Hewland transaxle. Suspension was all-independent, and brakes mounted outboard. Coolant circulated through radiators either side of the engine, just behind the doors.

For weight, *R&T* quoted 2030 pounds. However, Fouchee says that the No. 1 car with all fluids aboard save fuel (the tank held 31.7 U.S. gallons) scaled 1989 pounds—"right on the class limit." In overall length and width, the XJR-5s "were right out to the IMSA limits within an eighth of an inch"—just under 189 long and 79 wide. That's pretty big for a sports car—bigger in these dimensions than the notably bulky Ferrari Testarossa. But Fouchee points out that it was vital for GTP racers to occupy as much road area as the rules allowed to maximize potential aerodynamic downforce. The roofline was a diminutive 39.5 inches above the road with ground clearance set at the required minimum of 2.5 inches.

Rules determined not only the XJR-5's size but much about its shape. The most visible example was its rear wing. As IMSA's 1983 regulations specified that wings couldn't overlap the body outline as seen from above, "we put [the wing] where we wanted it and just ran the body out underneath," explains Fouchee.

Though handsome, the essentially wedge-shaped nosepiece was no visual match for the pretty prows on the old D-types and XJ13. Of course, aesthetics had less to do with this than aerodynamics. D-type body designer Malcolm Sayer had come from the aeronautics industry and worked in the age when straightline speed was prized above all, so he thought mainly of "penetration." His car was supposed to pierce the wind like a bullet. And it did. Despite fairly modest engine output, the racing D-type was very fast for its day. But though Sayer was aware of ground effects—the interaction between car bottom and road surface—to him it was an adverse phenomenon, a buildup of pressure beneath the car that could cause instability. That's apparently why he sharply slanted the D-type's lower body sides behind the front wheels: to aid the escape of air rammed under the car by the lower part of that bullet-nose.

Three decades later, Dykstra's XJR bodyshell reflected the realization that if ground effects could be made "negative," it would have a positive effect. The wedge-shaped nose not only set the air passing over it to creating downforce on the front wheels, but reached right down to that 2.5-inch ground clearance to keep excess air from building up underneath. The slab-sided bodywork behind the front wheels did a similar job, discouraging wayward swirls of outside air from degrading the low pressure being generated by the underbody.

Though seldom seen, the belly is really the most important part of a modern race car's body. When properly shaped, it can generate an astonishingly powerful "suction" at speed, first by trapping air as the car moves over it, then allowing it to expand in a carefully controlled manner. That expansion lowers pressure, which allows the higher pressure of outside air to push the body down. In a sense, the entire car becomes an inverted wing. The benefit, of course, is increased tire adhesion and higher cornering power. To keep speeds within the bounds of existing safety technology, governing bodies have long restricted the size and placement of wings, and now also severely limit the dimensions and positioning of downforce-producing underbody devices.

Negative ground effects in modern race cars is usually generated by a pair of venturi-shaped, open-bottomed recesses let into the floorpan either side of the cockpit, central fuel tank, engine, and transaxle. Commonly known as tunnels, from their appearance when viewed from the rear, these are typically about a foot wide and a foot high at the tail. However, toward the middle of the car the tunnels pinch down—sometimes right to zero depth, level with the otherwise flat floorpan.

That last describes the Group 44 Jaguar. Both IMSA and FIA mandated an absolutely flat bottom across the full width of the car under the cockpit, which obviously determined the point where the tunnels could begin. However, the sanctioning bodies had slight but significant differences in tunnel rules, and Group 44 had to choose between them. The American organization allowed a shorter floor, thus permitting longer, more effective tunnels. The Tullius team, however, built its Jaguars to the more restrictive European specs, deliberately accepting a handicap at home in order to optimize its chances at Le Mans. According to Fouchee, IMSA rivals running in France had to attach fill-in panels to the first few inches of their tunnels, which he says reduced their effectiveness to a surprising degree.

Curiously, but hearteningly, the single most

*American Bob Tuillius' Group 44 team took Jaguar's name back to the prototype sports-racing wars in 1982 with the 200-mph-plus XJR-5, which had a production-based sohc V-12.*

*To comply with rules prohibiting the aft wing from overlapping the body, the rear contour of early XJR-5s was shaped into something resembling tailfins.*

important element of the ultra-modern XJR-5 was something that had already made a successful crossover from road to track: the Jaguar V-12. As adapted by Group 44, the familiar old production powerplant proved competitive with some far more exotic racing engines, though in final form it was a long way from stock.

As Fouchee recounts, Group 44's first 1982 GTP engine was "the next generation past our Trans-Am engine." Over-bored by 0.040 inch, to 3.580 (90.9 mm) but with stroke at the stock 2.756 (70 mm), it displaced 332.9 cubic inches (5455 cc). With 12.8:1 pistons and six twin-barrel Weber carburetors, output was 560 bhp at 8000 rpm and 390 lbs/ft at 5500 rpm. In 1983, a further bore increase to 3.625 inches (92.1 mm) and a new crankshaft giving a stroke of 2.952 (75) upped swept volume to 365.6 cubic inches (5991 cc). Running with the Lucas fuel injection that eventually replaced carburetors, and with compression hiked to 13.1:1, Group 44's best-yet pumped out 631 bhp at 7500 and 496 lbs/ft at 5000.

The next displacement step was actually a two-step: two different bore/stroke combinations each equaling 6.5 liters. First came an impressive crankshaft with 3.200-inch (81.3-mm) stroke; installed in the 3.625-bore block, the result was 396.3 cubic inches (6494 cc). On dizzying 13.2:1 compression, torque was 555 at 6000, horsepower 688 at 7250. The second "396" employed the 2.952 stroke with new 3.770 (95.8) pistons for 395.4 cid (6480 cc). Optimum compression here was "only" 13.0:1, and torque fell slightly to 544 lbs/ft at 6250. But this version could wind higher, and proved capable of precisely 700 bhp at 7500 rpm.

In the last part of its last season, 1987, Group 44 was allowed to run 7.0-liter engines. Two were built, naturally blending the 3.77 bore with the 3.2 stroke for 428.7 cid/7024 cc. Compression stayed at 13.0:1, but torque went up to a mighty 596 lbs/ft at 6000 rpm and horsepower to 738 at 7250. Alas, this "stump-puller," as Fouchee calls it, ran in but two races. Other 1987 events were contested with either a 6.5 or 6.0.

Yet every one of the engines built throughout the six seasons of the Group 44 program retained the basic aluminum block of Jaguar's production V-12. Over time, of course, the team replaced most everything else with special hardware. The most impressive such development—and for long the most confidential—involved new, non-Jaguar cylinder heads.

At first, Group 44 used the original "flat-head" or Heron-type V-12 heads. They never considered switching to the later May "Fireball" design because its ports were too restrictive for racing. But the old Heron-head porting was not much more generous. As Fouchee explains: "The displacements we ran were larger than Jaguar's original intentions for that engine. On porting, we used to break through the intake ports in particular. There are sonic tests and other tests the head people use to show how thin the walls are, [but] they'd overheat and crack . . . We did have some success out of heads [Jaguar had] done for us. But when they changed to the May head . . . they scrapped the old patterns and had to recreate them for us, a year's work. We also needed the backs of the [new] heads squared up so we could drill and tap holes to take the rear suspension."

Tullius finally commissioned local foundries (first in California, later in Ohio) to produce two generations of special cylinder heads. Both basically followed Jaguar's original design, but had more meat around the intake and exhaust passages so they could be opened up to whatever dimensions were required for free breathing. In time, Jaguar itself began producing similar "thickwall" castings with bigger ports. These Heron-pattern retro heads were supplied both to Group 44 and to Tom Walkinshaw Racing, but never were used on production cars.

Though the stock cylinder block was always used, Group 44 did a lot of work on it. One problem concerned head gasketing. The Jaguar V-12 block is "open-decked," meaning its cylinders are separate, freestanding pieces mechanically secured only at the bottom. Torquing down the heads pinches the cylinders in place, and works perfectly well for street use. But when subjected to the rigors of racing, those 12 individual cylinders "dance all over the place," in Fouchee's fine phrase, resulting in a serious power loss as cylinder pressure escapes at the tops of the bores. After a great deal of experimenting, Group 44 achieved an effective solution with sealing rings made of small Inconel tubes filled with a salt compound. "The heat of welding it into a ring creates pressure," explains Fouchee, "so the hollow ring is resilient . . . That was the beauty of it. If the head moved, it moved with it."

Not even the block itself escaped the racers' restless revisionism. Lanky's records show that the first GTP engine weighed 627 pounds complete, except for exhaust system. Subsequent engines went on a diet. Some pounds were shed the old-fashioned way by simply machining off unneeded metal. Other pounds were lost by substitution: Titanium connecting rods weighing 50 percent less than steel took off 12 pounds; replacing the iron cylinders with Nickasil items saved 21; a carbon racing clutch dropped nine; and another eight were removed by swapping the stock General Motors alternator for a smaller Mitsubishi unit. Various covers were recast in magnesium instead of aluminum. The performance payoff from all this dollar and labor investment was Group 44's lightest-ever Jaguar V-12: 520 pounds—more than 100 pounds trimmer!

The XJR-5 first raced at Road Atlanta in 1982, when it actually won the GTP class while finishing third overall (the class was then still subsidiary to GTX, a province of turbocharged Porsche 935s). Into their first full season, 1983, XJR-5s still ran carbureted 5.5-liter engines to win four events outright, plus a second and a third. That was enough to put Bob Tullius second in the drivers' points championship and for Jaguar to rank third in the manufacturer's standings.

A promising beginning, but succeeding seasons brought only mixed results. For 1984, Group 44 enlarged the V-12 to 6.0 liters and fed it with fuel injection. That version managed to score one first in GTP, plus several seconds and thirds. But though much of the season's efforts focused on Le Mans, that too, proved disappointing, as both cars failed to finish the 24 hours.

Things went a little better in 1985. The 6.0 benefited from a new Lucas "Micos" electronic management system, and there were other improvements, such as a

*Group 44 followed the XJR-5 with the XJR-7 in 1985. It had a different tail shape and ran 6.0- and 6.5-liter naturally aspirated V-12s against mostly turbocharged competition. It was lighter than the XJR-5 and lighter than most rivals, but even with as much as 700 bhp, it didn't win its first race until '86 and won only two more in '87.*

rear suspension revised to suit radial tires, plus more refined aerodynamics. This XJR-5B (also called "Phase II") won two IMSA contests. Two cars were again fielded for Le Mans, where one lost a cylinder but managed thirteenth place overall and first in the special GTP class.

In all, Group 44 constructed 10 examples of its first-generation GTP racer. An eleventh "display" car was recently made functional, per commission from a Jaguar collector.

At the end of 1985, Group 44 introduced a new car, the XJR-7. Though similar in concept and appearance to the XJR-5, it had a lighter, stronger chassis tub making more use of composite and honeycomb materials, plus bulkheads rendered in aluminum rather than steel. The engine wore its starter motor at the front, and thus had to be connected to the flywheel ring gear with a long drive shaft. The flywheel itself was smaller, and some metal was machined off the bell housing. The purpose of all this was to narrow the entire engine package so the underbody venturi tunnels could be larger.

The XJR-7 didn't notch its first victory until the last IMSA race of 1986. It won two more the following year, which turned out to be Group 44's last with the Jaguar program. Over six seasons, the team had entered 76 events, including the two Le Mans contests, and actually made 120 starts. The big white-and-green Jaguars took nine IMSA races outright, plus two additional class victories for a win rate of just under 15 percent. They also recorded 17 seconds and 14 thirds.

Not at all a bad record, but no championships were won, and Tullius' hope for dominating Le Mans went unfulfilled. Unhappily, he never got a third crack at the French classic, though he'd planned a greatly revised third-generation car for the 1988 season based on one of the five XJR-7 chassis he'd built. Designed for newly liberalized rules, it was distinguished by a narrower, more aggressively streamlined "greenhouse" that allowed larger air ducts to the shoulder-mounted radiators. Cooling had been an inherent problem of the original body, especially as engine sizes grew. The car was finished but never tested, and never got an XJR model number.

A combination of politics and logistics was the reason. By mid 1985, Tullius had a direct competitor within the Jaguar family: Tom Walkinshaw, whose XJ-S racers had just beaten BMW for the 1984 European Touring Car Championship. The aggressive Scot soon convinced Jaguar that his Oxfordshire organization, so much closer to Coventry than the U.S. state of Virginia, was literally better placed to tackle not only Le Mans but the entire World Championship for Sports Racing Cars. And as Walkinshaw no doubt pointed out, this was to be run under FIA's Group C regulations, which were enough different from IMSA GTP rules to warrant building a wholly new car, one that could take advantage of the enormous strides in racing technology being made by various Formula One constructors headquartered nearby.

So Jaguar allowed its contract with Group 44 to expire. Tullius turned to other things, and fans of the ferocious felines again placed their hopes for a modern triumph at Le Mans with Coventry.

*Group 44 twice fielded a pair of XJR-5s* (below) *at Le Mans. Neither finished in '84, though one managed thirteenth place overall in '85. Tuillius planned a greatly revised car based on the XJR-7 for the '88 season, but Jaguar ended its ties to Group 44 and placed hopes for a Le Mans victory with England-based TWR.*

CHAPTER XIX

# REPRISE AT LE MANS

*This dynamically shaped billboard on the side of a French farmhouse chose the right marque to advertise the 1988 24 Hours of Le Mans as Jaguar captured its first victory at the Gallic circuit since 1957.*

Their sleek flanks smeared with the filth of more than 3300 racing miles, but their twelve headlamps outshining the afternoon sunlight, the trio of XJR-9LMs ran their final lap to congratulatory bouquets of exuberantly waving flags from every spectator enclosure and worker's post they passed.

"On the last lap we were going line astern down the Mulsanne at maybe 180 mph or more and I was astonished by the marshals. Not only were they standing on the track itself, waving their flags, but they were actually trying to touch the cars as they went past!" So recalled American Davy Jones, a Jaguar team driver in that mighty train, as quoted in England's *XJR* magazine.

Another, Irishman Derek Daly, told the specialist publication, "I have never taken part in such a high emotion event—for the last two laps I was waving back at everyone. It was a wonderful feeling, and so highly charged that some of the people in the pits were in tears . . ." One of those pitside, Eddie Hinckley, crew chief of the victorious car, said the things he'd remember about winning were "Listening to the [British] national anthem being played four times. And seeing Tom [Walkinshaw, the bluff Scottish team owner] with a tear in his eye."

Any race fan—and thanks to international television there were millions—who watched as Jaguar won its sixth *Vingt Quatre Heures du Mans* in 1988 must have been struck not only by how much had changed in the 31 years since the marque's last conquest of the Sarthe, but also by how little.

What has remained the same at Le Mans is the excitement of the racing itself, and the human emotion generated by winning. Jaguar's renewed commitment in the eighties to the world-renowned 24-hour event rested not only on its own glorious victories in the fifties, but on the five by Bentley in the twenties and thirties, and on Aston Martin's success in 1959. Even Ford of America's string of four triumphs in the sixties owed no small thanks to English design, construction, preparation, management, and driving skills. The endurance epic in France always has meant something special to the Brits, and even in years when a great British victory was hard to imagine, stalwart hoards of them still crossed the Channel every June to wave their Union Jacks.

What has changed over the decades, of course, is the technology involved, and not only that responsible for the enormously more sophisticated, speedier cars.

*Jan Lammers, co-driver of the winning XJR-9LM No. 2, flourishes the Union Jack amidst a throng of jubilant British fans celebrating the '88 triumph. A Jaguar driver said some track marshals were so overjoyed that on the last lap they reached out to touch his car as it screamed past at 180 mph. After 31 years of gradual decline and painful reconstruction, the Coventry Cat had clawed its way back to the pinnacle of sports-car racing.*

Back in the fifties, nobody in racing had even dreamed of on-board electronics, of computers plotting strategy in the pits, of live, full-color video broadcasts going out world-wide by satellite. There had been radio in the old days, of course, but in 1951 Coventry's little expedition to France had resorted to a hand-keyed telegraph system to convey news of the great British victory to the Queen. In 1988, Her Majesty could have tuned in her palace TV and ridden along with "her" drivers, gazing raptly over their shoulders as they scorched the Mulsanne Straight at more than 240 mph.

High technology has touched today's drivers themselves. On raceday morning back in 1953, William Lyons had tackled the physical fitness issue by pouring black coffee and brandy into Duncan Hamilton and Tony Rolt in relief of hangovers so they could drive for him—and win. In 1988, the Jaguar team strength of well over 100 people included official full-time professional athletic trainers and nutritionists. Before and during the race the team's 14 carefully conditioned drivers were scientifically fueled on a high-energy diet of yogurt, cereal, and pasta. Naturally they enjoyed quiet and comfortable motor homes in which to rest; whenever someone's turn came to drive, revealed *XJR*, he was "gently roused, blood pressure and pulse rates monitored again, and then he had a few minutes on an exercise bike to warm up." Before being allowed to strap into his four-mile-a-minute missile, continued the report, the driver had to engage a trainer in a short bout of shadow-boxing, "so as to check reactions and general alertness . . . no point in sending out a sleeping beauty." Such was the off-camera depth behind the image of waving flags, blazing lights, and spraying champagne.

Our own focus here, of course, is primarily on the cars, and how this generation of racing Jags glitters with fascinating detail technology.

The British-built XJR racers from Tom Walkinshaw Racing (TWR) in many ways were similar to those created in America by Bob Tullius' Group 44, but the apparent commonality owed little to any copying. In fact, there was not much cooperation between the rival crews. Generally, any duplications more likely reflected similar logical solutions to the same engineering problems. And there were in fact numerous differences between the cars. Many of them grew not only from diverse approaches and resources, but also to variations in regulations between the U.S. and European sanctioning bodies.

FIA Group C, inaugurated in 1982, called for vehicles basically of the same type and design as America's somewhat older IMSA GTPs, but neither kind of car was eligible for the other's series without considerable detail restructuring. In the previous chapter we noted that the rules governing the Euro-car's ground-effects-generating underbody (sometimes called the "underwing") were

*TWR's first Group C car, the XJR-6, bowed in '85. It ran a 6.2-liter sohc V-12 under British Racing Green* (right). *Note ground-effects-enhancing rear-wheel spats. Revised for '86 with a 6.5 V-12 and new paintwork* (below), *it won at Silverstone, but couldn't finish Le Mans.*

more restrictive. On the other hand, the FIA allowed a significantly more effective rear wing earlier than did IMSA. The Group C car also was allowed to weigh 110 pounds less—850 kilograms (1874.2 pounds) versus the IMSA minimum of 900 (1984.5). European engines were essentially free of limitation, too, so they could be both large and turbocharged. American engine rules were not only more restrictive, but subject to constant review by an IMSA concerned with inter-make parity.

FIA addressed the same consideration through a different fuel philosophy. Whereas in IMSA gasoline quantity and quality were essentially free (though onboard supplies were held to 120 liters or 31.7 U.S. gallons), Group C cars could carry only 100 liters (26.4 gallons), and the fuel octane was limited to 102. Moreover, consumption could be no more than 51 liters per 100 kilometers of race distance (4.6 miles per gallon). The need to conserve fuel automatically restricted performance. Thus, except during qualifying or brief sprints during races, when a turbocharged engine's boost pressure might be raised temporarily without harming the overall fuel picture, drivers of the relatively unlimited FIA engines enjoyed little, if any, more horsepower than did their IMSA counterparts.

As we have seen, Walkinshaw founded his excellent relationship with Jaguar by winning the 1984 European Touring Car Championship with his team of race-prepared XJ-S coupes. The forcefully ambitious "Major Tom" then talked Coventry into helping him move up to Group C, virtually in direct competition with Tullius. His Kidlington, Oxfordshire, shop started laying out its first pure-blood racer in November, 1984. Since that predated Group 44's second-generation GTP, the British car received the designation of XJR-6, leaving -7 to the American model. Initially, Walkinshaw's men spurred themselves on with hopes of making a run at Le Mans the next June.

Primarily the work of Tony Southgate, an English designer with credentials not only in Group C but also in Formula One and Indianapolis, the TWR machine was similar in basic layout to Dykstra's for Group 44. Conceived from the outset as another closed coupe, and mid-engined, of course, it too used the massive Jaguar V-12 as a stressed chassis member with only a quartet of extra bracing tubes. Also, per regulation it carried its fuel in the middle, between cockpit and engine.

However, Southgate differed from Dykstra by giving the TWR car a "wet nose"—a single large radiator right up front. Perhaps he'd learned something from Group 44's progressive troubles with overheating in its

*Jaguar returned in '87 with the XJR-8. Stroked to 7.0 liters, the V-12 made about 720 bhp, 20 more than the 6.5. It's shown here in Le Mans garb. This XJR-8LM eliminated the rear-wheel spats to ease tire changes. That cost some speed and downforce, but aero refinements elsewhere more than made up for the losses.*

mid-mounted cooling system as engine output kept rising, but he also wanted as much weight carried forward as possible.

Also, the TWR man was working within more liberal FIA rules then governing cockpit dimensions. That enabled him to give the XJR-6 a narrower, aerodynamically smoother canopy. Concern for clean airflow also led to enclosing the rear wheels in "spats" eerily reminiscent of those on the original XK 120 street car of 1948. Those on the '85 racer had nothing to do with style, though, as in fact a great number of hours of wind tunnel testing of one-fifth-scale models went into the new body.

That body was blunt-tailed and had a startlingly unconventional look. Obviously, after all this testing, it had proven itself aerodynamically correct, and the clue lay in the fact that the visible part of the shape was not as important as the hidden. As Southgate confirmed in a book, *Jaguar XJR*, by Ian Bamsey, "The general body shape was designed primarily to benefit the operation of the underwing and the rear wing." In other words, the top of the body served mainly to channel air smoothly and efficiently to the rear wing; the wing's primary role was to create low pressure behind the car, to forcefully draw air out of the underbody venturi tunnels (also known as engine bay diffusers), thus increasing downforce.

Indeed, so important did the designer consider underbody aerodynamics that, in an apparent paradox, he made the basic car narrower than the maximum allowed by the rules. For Group C, the maximum overall width was 2.0 meters (78.7 inches). In the modern ground-effects era a wider car is a better-cornering car, but his extensive testing had shown Southgate that shelf-like extensions jutting out from the bottoms of the body sides would block some exterior air from being sucked into the underbody diffusers, thus preserving their effectiveness. So he shrank the bulk of the XJR-6 body to 1.9 meters (74.8 inches), simply to permit these small shelves. Of course, a narrower car should be a faster car in a straight line. Southgate had achieved one of those rare tradeoffs that racing engineers stay awake dreaming about: he'd gained something without actually paying for it.

Similarly, the new Jag's other dimensions derived less from pure science than regulatory directive. As for overall height, the rules set that at 1030 mm, or 40.6 inches. Maximum permissible overall body length was 4.8 meters (188.9 inches) including the rear wing; fair enough, but there was another clause controlling how far the bodywork could extend ahead of and behind the wheels. The actual rule specified that the total extension, front plus rear, could not exceed 80 percent of the wheelbase dimension. Confusing, but it mattered, because the front and rear overhangs of modern race cars have profound effects on aerodynamics. However, the long Jaguar engine forced the XJR-6 wheelbase up to 2780 mm, or 109.4 inches—leaving a good eight inches of permissible overhang length unused.

We may, for a moment, take a breath and speculate as to what words might have been selected by the designers of the so small, so simple, so elegantly straightforward Jaguar C- and D-types to express their opinions about all this latter-day regulatory obfuscation.

But racing science marches on, and Tony Southgate managed to find plenty of room within the rules to exercise his creative talents. In designing the rest of his car, he took full advantage of one of Walkinshaw's points of argument to Jaguar, the ready availability in England of the most advanced Formula One construction knowhow. According to the manufacturer's publicity, the TWR machine was the first sports-racing car of the modern era to have an "all-plastic" chassis tub. The materials actually used were sophisticated aerospace composites, such as carbon fiber, Nomex, and Kevlar. In fact, the structure still contained some metallic elements, such as aluminum honeycomb in certain paneling and protective steel tubing in the canopy. Nor should we forget that experimental, unraced D-type apparently made mainly of fiberglass by Jaguar itself in the fifties, as well as several of the 'glass-tubbed American Chaparrals that raced very effectively in the sixties.

Propaganda aside, the XJR-6's was an immensely strong and stiff yet extraordinarily light form of construction, and certainly was a forward step. We can admire, especially, the way the designer seized an opportunity suggested by the technique of actually molding the chassis into shape. For proper handling balance, Southgate wanted to carry the weight and bulk of the engine as far forward as possible. At the same time, safety regulations required the driver's feet to remain behind the front axle line, and also that the fuel tank be sheltered in the middle of the car, behind the cockpit.

Granted, the FIA-legal tank was 17 percent smaller than IMSA's, but to cram everything together there would have to be a sort of hole in the fuel bay for the engine. Building an intricate, wrap-around module tightly surrounding the front of the V-12 would have been a difficult job of fabrication using metal, but composite layups made it almost easy.

In planning its Group C engine, TWR had been free to turbocharge the Jaguar V-12, but the team's powerplant specialist, Allan Scott, decided the performance/

*The XJR-8 was Jaguar's most successful modern-day sports-racer yet. TWR won the world sports car title with it in '87, and campaigned three XJR-8LMs in that year's 24 Hours of Le Mans* (both pages).

economy equation could best be satisfied by keeping it naturally aspirated. Another option was a new cylinder-head design, and TWR used one of Group 44's XJR-5s as a test mule to evaluate a four-valve-per-cylinder engine. Though somewhat more powerful and easier on fuel, the four-cam, 48-valve configuration added substantially more weight to a "lump" that the chassis designer already considered too heavy and bulky. And Scott remained enthusiastic about the ability of Jaguar's basic single-cam, two-parallel-valve layout to efficiently flow air and burn fuel. Just as with the American team, too, TWR's several years of successful XJ-S racing had built confidence in the "old" design. So TWR's XJR-6 emerged with an engine that looked very much like Group 44's.

Appearances deceived, because the actual heads used were something special inside. Something secretly special. TWR has always refused to reveal details of its combustion chamber, porting, and pressure-sealing design.

But most of the other information involved is known. The 1985 Group C engine had bore and stroke dimensions of 92.0×78.0 mm (3.622 by 3.071 inches) for a swept volume of 6222 cc (379.7 cid). The block came from the Jaguar production line, though Scott had it remachined in some areas, and the new crankshaft was held in place with special four-bolt steel main bearing caps much stronger than stock. As on the B-series Group 44 XJR-5, the TWR car gained space for its venturi tunnels by relocating its starter motor toward the front of the engine. Weight of the engine, including the three-plate clutch pack, was said to be 250 kgs. That's a suspiciously even figure, but for what it may be worth, it converts to 551 pounds. The electronic engine management system, made by Zytek, was derived from Formula One practice. On a compression ratio of 12.0:1, the quoted maximum horsepower was 660 at 7000 rpm. Redline was set at 7600.

So far these are rather familiar Jaguar GTP figures, and when the first two Group C cars appeared for their inaugural race their actual weights were IMSA-like too: 910 and 895 kg (2007 and 1973 pounds) respectively, well over the FIA minimum.

That first race was not Le Mans. By the time all the exquisitely crafted bits—many from outside suppliers and specialist fabricators—had come together for the first test, it was July. But the pair of XJR-6s were ready in August to go to Canada for an international race at Mosport Park. The brand new British racers, painted "British Racing Green" at that stage, by no means disgraced themselves in the New World. One qualified third to a pair of turbocharged Porsche 962s, and actually led

*One XJR-8LM reached 244 mph on the Mulsanne Straight in '87, but of the three entered, one crashed, the second broke, and the third finished a frustrating fifth.*

the race for 10 laps before retiring with a failed wheel bearing. The second Jaguar finished in third place running on 11 cylinders.

There were two more Group C races in 1985. One of the new Jaguars finished merely fifth in one, but took a competitive second in the finale. No wins, but the overall performance was encouraging enough to let Jaguar and TWR plunge into a winter of contentment as they revamped and refined a definitely promising basic design.

A diet was the first order of business. The three 1985 cars built had been built heavy. Southgate ran through some of his structural calculations again, and found places where inexperience with the new materials had led him to be too conservative. By using lighter-gauge composites in some areas of the tub, by stepping back to simple fiberglass instead of more exotic stuffs in the bodywork, and even by substituting aluminum parts for composites here and there, he managed to save about 20 pounds. The engine people gave him another 20 or so; the new engine-weight figure was 240 kg, or 529 pounds. When finally screwed together, the two new 1986 chassis—still called XJR-6s—scaled a tidy 851 and 855 kg (1876 and 1885 pounds) for short-distance, or "sprint," racing. With certain extra equipment necessary for the Le Mans enduro, they came up to about 870 kg (1918 pounds).

Jaguar would, of course, make it to Le Mans this time, and with a special low-drag/low-downforce version of the basic aerodynamic package. When clad in this subtly revised bodywork and wing system, the car was called the XJR-6LM.

Nor did the work stop there; mechanical changes for '86 were legion. Increasing cylinder bore to 94 mm (3.701 inches) pumped displacement up to 6495.6 cc (396.4 cubic inches) and, with other refinements, brought horsepower to about 700 at 7300 rpm. The brawnier engine breathed through a new rooftop airscoop. Less easily seen improvements extended throughout the car: in the electronic engine controls, the March-built five-speed transmission, the brakes and suspension, even the fuel tankage.

On a larger scale, TWR and Jaguar fine-tuned their very methods of going racing together. In Coventry, the factory's new engineering center at suburban Whitley was in business, and could contribute a broad spectrum of services, from stress-testing suspension components to bedding-in new brake pads. Down at Kidlington, which now was geared up to make most parts of the cars in-house, every highly stressed component was assigned a serial number so a computer could keep track of its service life.

*Jaguar was back at Le Mans in '88 with the XJR-9LM, its sohc 7.0 V-12 up to 740 bhp for European tracks. The car was more hospitable for drivers, including Johnny Dumfries* (middle in photo below) *and Lammers* (right).

*After victories at Daytona Beach, Monza, and Silverstone, Jaguar descended upon Le Mans with five XJR-9LMs, 14 drivers, and a total crew of 110. Sharing the field were 11 Porsches, four Nissans, and two cars each from Toyota and the French team WM. Tire woes scratched the two Sauber-Mercedes entries.*

Painted a bold white, gold, and purple color scheme in honor of new sponsorship, the second-generation XJR-6 launched the season at Monza, Italy, where it showed competitive speed against the Porsche, Sauber-Mercedes, and Lancia entries, though it failed to finish the 221-mile sprint. But the second round of the series, a 1000-kilometer (621-mile) semi-enduro, resulted in victory at last. The venue was Silverstone, Britain's ancient airfield course that had helped make the XK 120's reputation 37 years before. Jaguar's latest success marked its first in international sports-car competition since Le Mans 1957.

At Le Mans 1986, one of the trio of LM-bodied XJR-6s entered was aerodynamically tuned for straightaway speed and was clocked at a best of 221 mph. An impressive figure, and nearly 50 mph better than the old D-type, once the fastest of its genre. However, the -6 was fully 11 mph slower than a Porsche. Bummer? Not necessarily. All of these speeds were put into perspective by the fact that another of the XJR-6LMs, obviously tuned more for downforce, was able to show "only" 212 mph on the straight—yet it set a quicker overall lap time than its "faster" sister. The point was clear: even on the ultra-long Mulsanne, flat-out velocity was no longer a priority, and using such a number to judge a design was meaningless. A car's top speed now was nothing inherent, but really just a matter of how its crew cared to set it up.

In their first Le Mans, the XJR-6s ran encouragingly against the Porsches, but none lasted the 24 hours. Essentially, that was the story during the remaining nine events of the season: good placings, but no wins.

TWR approached 1987 with its design revised in so many ways—Jaguar publicity enumerated 64 "significant changes"—that the model designation was upped to XJR-8. Three new cars had been constructed, all to the same basic configuration although the chassis material was now aluminum honeycomb rather than composites. On the other hand, the bodywork went back to the original Kevlar. It still looked much the same, though it was cleaned up aerodynamically in some areas, and to comply with new safety rules the doors were now more conventionally hinged at the front, rather than gull-winged as before. Also, to quicken wheel-changing the spats disappeared from the rear bodywork for 1987. By itself that robbed the car of a little bit of speed and downforce, both, but refinements elsewhere more than made up for the losses.

Under the new skin, the rest of those 64 improvements included a new generation of Dunlop racing tires, still bias-ply but with the plies made of Kevlar fabric. Like radials, they didn't expand with centrifugal force, so at

*The Jaguars were running over 240 mph on the Mulsanne, but a WM was clocked above 250 and the three turbo-Porsches qualified faster to grid ahead of the XJR-9LMs. But endurance races seldom are won on pure speed or starting position.*

high speed they maintained the low ride height vital to a ground-effects car. Watching over the health of the new tires was a "Heat Spy," an infra-red temperature monitoring system mounted in each wheelwell. Among many other details, enough additional weight had been saved that, in sprint trim, the cars had to carry some ballast to make the 850-kg weight limit. However, in Le Mans form the weight had crept up another ten kilos to about 880 (1940 pounds).

In fact, the XJR-8LM configuration was even more specialized than before, the product of a dedicated subteam assigned by TWR just to that job. To address a problem with the joints of the halfshafts running out to the rear wheels—a problem experienced about that time by Indianapolis cars also running March transmissions—the enduro Jaguars had the rear of their engine/gearbox package raised. That flattened the shaft angles and made the hard racing life easier on their joints, at the expense of raising the engine center of gravity about an inch. This doesn't sound like much, but it was plenty to the drivers. As they hurled the stiffly suspended machine into corners, they complained that the -8LM felt like it was trying to pitch its engine over their shoulders. Their bosses basically retorted that lasting 24 hours was more important than enjoying it.

Besides, the drivers had even bigger engines to enjoy this year. A longer stroke of 84 mm (3.31 inches) bumped displacement up to 6995.3 cc (426.9 cubic inches). The main benefit was not top-end horsepower, which increased by only about 20, so much as mid-range torque. The previous year's 6.5 engines had been rated at some 570 pounds/feet, whereas the 7.0s churned out 605 lbs/ft at 5250 rpm. The entire torque curve was very flat, too, making for an easier engine to drive, one that required less shifting, and also one that could be driven generally at lower rpm for better economy.

Fuel consumption still was important, although non-turbo engines such as Jaguar's benefited by a regulation change for '87. Earlier, the nominal 102-octane limit had been met by clever chemists who used extreme amounts of toluene to mix up a "jungle juice" that was denser and more resistant to detonation than everyday gasoline. These characteristics especially suited turbocharged engines, and Walkinshaw cried foul. Reverting to normal "pump" gas for the new season cut down everybody, but Jaguar less than the turbo-running teams. In fact, despite the effectively "poorer" fuel, the V-12's compression ratio was now up to 12.8:1.

All of this makes for a tediously technological tale, perhaps, but the outcome was terrific: nothing less

*One Jaguar collided with a Porsche during the race. It was fixed and continued on. Three XJR-9LMs had engine and gearbox troubles; two dropped out, another had a long delay. The surviving cats roared round the clock.*

than eight Jaguar victories in the 1987 season's 10 races, and the FIA World Sportscar Championship of Teams. One of the pilots, Brazil's Raoul Boesel, earned the driver's world sports car championship.

Alas, the short loss column included Le Mans. Once again the Jaguars showed plenty of speed—up to 227 mph on the Straight (one -8LM driver reported seeing a stunning 244)—but less reliability. In the race, one car blew a tire at maximum speed and suffered a mighty crash; happily, the chassis structure proved immensely strong and the driver walked away—albeit in a wobbly fashion, one must suppose. Another Jag went out with a cracked cylinder head. The third, too, was beset by mechanical problems, though it managed to finish fifth.

Whew. This Le Mans stuff was tough. There might have been one or two old Coventry hands who caught themselves longing for the good old days, when racing seemed simpler. But competitive spirits love racing precisely because the challenge is endless. Everybody involved plunged straight into another cold English winter's worth of work.

Once again the FIA had presented a list of revised regulations, and now in 1988 TWR had to cope with IMSA rules as well. Having taken Jaguar's North American flag from Group 44, Walkinshaw set up a new base at Valparaiso, Indiana, a university town in the northwest corner of the state, not far from Chicago. To simplify logistics, he directed Southgate to provide the U.S. team with cars that had as many parts in common as possible with the European version. That called for some compromise.

The '88 Jaguar once again grew out of the previous edition, but had enough changes to warrant a redesignation to XJR-9. Both the Group C and GTP models were indeed very similar and for Le Mans, two IMSA cars would temporarily be converted to FIA spec.

Wheel sizes were the major fundamental structural difference between the rules of the two sanctioning bodies: IMSA required front and rear rim diameters to be the same. The FIA Jags had been running 17-inch front, 19-inch rears. Going larger at the front didn't make sense, so to achieve commonality Southgate had to redesign his basic rear suspension and bodywork to take smaller wheels and tires. That intruded into the space formerly given to the underbody diffuser tunnels. But he was glad to let it. For one thing, by coincidence the Group C rules had just cut that space anyway, to reduce downforce. No loss there. But smaller rear wheels meant a lower body line, which reduced drag (to the same end, he put the rear wheel spats back on for '88). The biggest windfall of smaller wheels, though, was the opportunity to drop the engine in the LM version back to the original design height without hurting the half shafts. And lowering the engine put a stop to all the moaning from the drivers about handling. In the great, endless tradeoff tournament that is race-car engineering, Tony Southgate had won another round.

There was one other significant structural edict, this one from the FIA: For 1988 the "flat-plate" area of the bottom of the car had to be larger, 900 mm long and stretching the full 2000-mm width of the car. So Southgate had to revise the composite chassis a little anyway, and while he was at it he moved the front wall of the centrally located fuel tank forward a bit. That increased the interior volume so it was ready to accept the extra 20 liters IMSA allowed. But these changes were so modest that, in some cases, the crew crafted "new" cars out of older ones.

They also found the rest of the FIA/IMSA discrepancies easy to handle—adding ballast to meet the extra 50 kilos of weight required by IMSA, fitting a different underbody with the larger tunnels allowed by IMSA (they still managed to fit the space allowed by the 17-inch rear wheels), changing to softer springs and gas-charged shocks to suit the typically rougher North American tracks, beefing-up the gearbox to cope with the same factor. Eventually, the IMSA cars adopted Formula One-type carbon-on-carbon brakes to handle the generally tighter Stateside turns. At times, too, the GTPs ran radial-ply tires.

Engine-wise, the European displacement remained at 7.0 liters, though various upgrades improved horsepower to as much as 740. Even more—up to 830 horsepower according to some reports—was extracted from a second-generation twin-cam, four-valve head design. Much more efficient than the 48-valver tried before, this one's weight penalty over the single-cam version came to only about 17 pounds. TWR tried the exotic powerhouse in one race early in the season, but it experienced mixed results and dropped the project.

In the U.S., where Group 44 had briefly run its own 7.0 engine the year before, new 1988 rules cut the Jaguar back to 6.0 liters. This was easily accomplished with the familiar old 94×72 mm dimensions, and by running an elevated compression ratio of 13.5:1 to take advantage of IMSA's freer-fuel formula, TWR managed to see 670 horses at 7500 rpm. The torque reading was 535 lbs/ft at 6250.

Both versions of the XJR-9 were equipped with a new transaxle casing that was produced in-house by TWR, although it retained March five-speed gearsets. There were many other cleanups all around the cars to

*"Our first task was to get through the night safely," Tom Walkinshaw explained. "When we'd done that, we turned the pressure up a bit . . . up a bit . . . and when the Porsche drivers wanted to ease back in their normal way, we just wouldn't let them do it."*

improve functionality; the drivers even got electrically adjustable mirrors!

On the LM model variation, enough cleaning-up was done to get the weight down to 855 kg (1885 pounds). As before, the LM body was easily distinguishable by its special nosepiece, which featured quadruple headlights but lacked pressure-relief louvers atop the front fenders. A closer look revealed no "splitter" jutting forward from the bottom of the nose, and that the rear wing was mounted lower behind the tail. The underbody tunnels were designed to create less "suction," too. Altogether, compared to the sprint-race configuration, in its Le Mans guise the XJR-9 was said to experience no more than half the aerodynamic download at any given speed. The payback, of course, was reduced aerodynamic drag.

The North American season got off to a rousing start at Daytona Beach in January, when Jaguar finally broke a Porsche streak of 11 straight victories in the 24-hour GTP race there. A bit less rousing—but only a bit—was the beginning of the European season, where in March, at Jerez in Spain, the Group C team took second place to a Sauber-Mercedes. But the following weekend, at Madrid's Jarama circuit, TWR managed to turn the tables for a win. The Jaguars went on to take Monza too, then scored a repeat victory at Silverstone. That set the team up nicely for the assault on Le Mans.

And a major assault it was. The IMSA team contributed two cars converted to FIA rules, for a total strength of five XJR-9LM race vehicles. There also was a spare tucked away in a trailer. The TWR personnel roster, including 14 drivers, came to 110. And that didn't count additional people providing services, such as catering. Arrayed against the British army were no fewer than 11 Porsches, four Nissans, and a pair of racers each from Toyota and the French WM team. Sauber-Mercedes loomed as a formidable threat at first, but tire trouble in practice forced it to scratch its two entries.

That problem with tires also affected the Jaguars; it was to a lesser extent, but it gives us a particularly telling insight into the complexity of the modern racing machine. For 1988 the long Mulsanne Straight had been repaved. Fine, but the new surface was so much smoother that it allowed the cars to run lower to the surface for better handling and more effective ground effects. Fine again—except that, even dressed in its low-downforce body, the XJR-9LM was tearing along the straight so fast that it was generating more "squash" than ever before. The poor tires were running so hot their treads were blistering. Unlike Sauber, which had one explode, TWR was able to moderate the heat buildup with higher tire pres-

*Jaguar chairman Sir John Egan had to leave the TWR pits, so great was the tension. "I thought the whole race typified Porsche's incredibly competitive spirit," he said. "They* never *gave up. . . ."*

*The all-conquering XJR-9LM of Lammers, Dumfries, and Andy Wallace is engulfed at the winner's circle. It beat the second-place Porsche by less than three minutes. The two other surviving Jaguars finished fourth and 16th.*

sures and go on.

How fast were the latest Jags on Mulsanne? One driver's tach reading suggested 244 mph and another source indicated 241, though the car's own designer believed such heady velocities were a bit optimistic. As before, of course, pure speed wasn't the main issue, neither on the straight nor even, in this case, around the rest of the lap. TWR was not concerned that a WM was clocked at better than the magic 250 mph in a straight line. Nor was it troubled that three turbo-Porsches (highest claimed top speed: 243) turned quicker overall lap times than the best Jaguar in qualifying, relegating the cat to fourth on the grid. Particularly in an event governed by fuel consumption, starting position would have nothing to do with the finishing position 24 hours hence.

Endurance racing has accurately been likened to war, and Le Mans '88 was a real one. Once again, the main combatants shook out as Jaguar and Porsche, and they joined battle right from the start. The fastest of the German cars preserved its starting advantage for several laps, but one of the British team's fastest drivers, Dutchman Jan Lammers, soon stormed up from his sixth starting position to take over the lead. But then one of the other Porsches bashed the leading Jaguar in the tail, cracking the bodywork. Replacing that would have caused a crippling delay, so the Walkinshaw mechanics hastily patched it instead. In later hours, the same XJR-9LM was hit by a rock, which split its windshield. That did have to be replaced, but again the inspired crewmen did the job with amazing speed.

The war blazed on into the night, on into the misty dawn, on into a mid-morning rainstorm. One of the threatening Porsches ran out of gas, losing two crucial laps, and some of its fellows ran into other problems. But engine and gearbox trouble struck three of the Jaguars; two of the -9LMs had to drop out entirely, and another had a very long delay. There also were trivial incidents that made hearts pound: brake pads and wheel hubs that stuck during pit stops, drivers who "put a wheel wrong" out on the course. But the surviving cars soldiered on. The Jaguar men were absolutely determined to win this time.

"Our first task was to get through the night safely," Tom Walkinshaw explained to *XJR* magazine. "When we'd done that, we turned the pressure up a bit . . . up a bit . . . and when the Porsche drivers wanted to ease back in their normal way, we just wouldn't let them do it.

"There were two battles going on, really: the

physical battle between the cars, and the psychological battle between the team management on either side," Major Tom said. "We knew all along that we had to push the Porsches hard. The only way to win would be to *race* them."

For Jaguar chairman Sir John Egan, watching from the TWR pits, the strain was agony. "I had to go away for a while, the tension was so great," said the mastermind of Jaguar's return to competitive stature both on and off the track. "I thought the whole race typified Porsche's incredibly competitive spirit. They *never* gave up: they were attacking all the time."

But not hard enough, in the end, to beat the Brits. Co-driven by Lammers, Johnny Dumfries and Andy Wallace, and leading its two surviving teammates in fourth and 16th places, the winning Jaguar XJR-9LM No. 2 swept past the checkered flag after covering 394 laps, 3313.64 miles or 5334.96 kilometers, in 24 hours, three minutes, eight-and-a-quarter seconds of racing. Its victory margin over the second-place Porsche was less than three minutes. A great triumph in itself, breaking as it did the seven-year domination of Porsche's Panzer at Le Mans. But to the teary-eyed, hoarse-throated, flag-waving English race fans it was much more: proof positive that, after 31 years of gradual decline and slow, painful reconstruction, the Coventry Cat had clawed its way back to the pinnacle of the highest mountain in the sports car world.

Of course racing, like life, goes on, and within days the TWR team was girding for the next battle. Porsche effectively dropped out after Le Mans, so the main opponents for the rest of the 1988 season were the turbocharged Mercedes-powered Saubers. Jaguar beat them only twice in the remaining six races, but that made a season total of six victories and another Teams title; this time Martin Brundle took the driver's championship. On the North American IMSA front, where they faced the all-powerful turbocharged Nissans, Jaguar's Daytona-winning GTP cars wound up the 14-round series with only one additional win.

As it went into 1989, 18 long years after its birth as a passenger-car engine and fully seven after finding itself reborn into top-line sports-racing competition, Jaguar's wonderful old single-cam V-12 was finally showing its age. There was racing life left in the old girl—in fact, there was one more Le Mans victory left—but already she was passing the technological baton to a new generation of Jaguar.

*After years of faithfully crossing the Channel to support favorites who had little chance of winning, British fans finally had a victor to cheer in that June of '88. From the tears and sweat and fatigue emerged a certainty: Jaguar was back!*

CHAPTER XX

# BREEDING TOMORROW'S JAGUARS

Jaguar's 1988 triumph at Le Mans, the more significant because it was so hard-won, sent a charge of euphoria and pride through the resurgent company. Victory on the track is undeniable—and undying; the good it does lives long after the deed. But while some automakers might have pulled back lest a subsequent loss tarnish their trophies, Jaguar earned further credit with its fans by battling on.

Still, even the most performance-minded automaker knows that victory on the track is not enough. The prime requirement always is to meet the interests, needs, and tastes of those racing success brings into the showrooms. So even while Tom Walkinshaw girded his loins in Kidlington to carry Jaguar's banner back into combat around the world, Browns Lane was going forward with future production models.

Or was it? Jaguar might have been moving at racing speed on the track, but there was little sign of similar pace in the design studio.

After dreary years of subservience to Leyland,

Jaguar was free and independent again, was racing again, and was again feeling its blood running hot. It also had regained at least some of its old aura with the public and press. Yet despite renewal of this wonderful winning spirit, nothing new or exciting to capitalize on it seemed imminent.

Dealers were clamoring. The XJ-12 was a not-so-sweet 16 by now, and a replacement based on the just-announced new XJ-6 (XJ40) was needed, or so they said. Equally desired was an XJ-6 "limousine" with a longer wheelbase, and there was some interest in an Estate Car—a station wagon. Some were even starting to see the need for an entirely new XJ-6 platform. And there was still grousing about the aging XJ-S. Going topless had perked up its image, but that was no substitute for genuine youth. Many dealers also wanted a smaller, more sporting Jaguar "saloon," a sort of neo-Mark II, which they thought of as a BMW-beater. There would eventually have to be a new family of engines, too: smaller, lighter and more fuel-efficient. Jaguar's new Whitley research and development center was known to be exploring all these projects, if not actually developing them, but there was no hope that any would see production within anything like the immediate future.

Most glaring, at least to some eyes, was the perpetual absence of a Jaguar about which nobody on the outside knew very much, but which everybody knew as the "F-type." Yes, a new sports car. Since the weary old Series 3 XK-E had been quietly laid to rest at the end of 1974—not "replaced" by the much bulkier, heavier XJ-S—there had been no truly sporting vehicle in Jaguar's lineup. This was a situation that many observers and most marque partisans felt must surely be put right. Granted, such a specialized machine would likely have sold in far fewer numbers than the XJ-S coupe and convertible, not to mention the four-door XJ-6. But William Lyons, creator of the SS Jaguar 100 and the XK 120 "Super Sport," had known how important a high-performance flagship was to his entire fleet. It was not only the kind of car that every enthusiast dreamed about, it was the kind that brought those same enthusiasts into showrooms, even though they might leave in a less *pur sang* cat.

*Jaguar bounds toward the 21st Century with not one but two supercars. Both are mid-engine two-seaters. Reviving the dual-purpose race-and-road-car ethic is the V-12 XJR-15* (opposite). *The more-luxurious XJ220—pictured above in its original V-12 form—now uses a twin-turbo V-6.*

The very presence of the posh, grand-touring XJ-S fueled anticipation of a more sporting conveyance, an F-type. But as year after year went by and the seventies became the eighties, the only thing that appeared was more rumors. This hardly discouraged enthusiast magazines from running periodic speculative stories on what was invariably termed a "still-secret" project. Most were naturally accompanied by imaginative, sometimes oft-repeated illustrations of the purported styling, which was deduced—or wished—to be an evolution of the immortal E-type's "hollowpoint bullet."

Still, public interest was clearly there—in fact, stronger than ever by the late eighties after some 14 years of no-shows. What's more, Jaguar's revitalized marketing thrust *needed* a bold and beautiful new two-seater F-type in the lead.

So where was it? Mired in a slough of slack funding, muddy design concepts, and tangled government regulations.

In fairness, creating automobiles had become enormously more difficult since the XK 120 days, when Bill Lyons and a couple of his boys could cut down an old chassis and hammer up a pretty prototype in a matter of weeks. Though the E-type, in its day the most technically sophisticated sporting Jaguar ever, had taken several years to reach production, those responsible always had clear sight of what they were doing: turning a racer, the D-type, into a road car.

But a quarter-century later, raceway and roadway had diverged so widely in their requisites that one could offer no path to the other. Jaguar product planners, already bereft of Sir William's brilliant leadership in matters of style and commerce, were laboring under exponentially increased costs. They were stumbling over continually shifting grounds of public whim. And they were fighting rules and restrictions of an unprecedented complexity and capriciousness.

Given the problems, the only feasible way to build a new, low-volume sports Jaguar was to erect its body on the chassis of an existing model. Had the groping firm only succeeded in coming full circle? That was precisely the way Swallow Coach Building Company had started out in 1927!

In 1978 came an offer of guidance from Italy, another nation of bubbling automotive enthusiasm. Pininfarina, the great *carrozzeria*, was known primarily for its many beautiful Ferrari bodies, but was also responsible for a number of successful British projects. The popular, low-cost MGB sports car of the sixties and seventies was one such effort, as was the crisp new roofline being penned for Jaguar's own 1979 Series III XJ-6. Hoping to inspire future commissions, the Turin coachbuilder asked to take over a worn-out XJ-S development hack, proposing to chop its coupe top, strip remaining outer panels, stiffen the naked structure, and then clothe it with a new open bodyshell.

Dubbed "Pininfarina XJ Spyder," the result debuted at the 1978 British national auto exhibition at Birmingham, where it created vast excitement. A Targa-style two-seat roadster, it was dominated by a long, rounded nose with oval air inlet, again strongly reminiscent of the late, beloved E-type. Wind-tunnel tests showed a drag coefficient of 0.36. That wasn't outstandingly low even then, but it was ten percent less than the 0.40 of the standard XJ-S Coupe (though PF never got around to fitting door mirrors). Other attractions included a prominent center console that made for snug individual cockpits per Jaguar custom, an automatic-transmission shifter recessed within the console as an alleged safety measure, and a "black-out" instrument cluster that lit up in a kaleidoscopic digital display when the ignition key was turned. Eyes perceptive of practicalities noted a generously long, flat baggage bay.

If Pininfarina's work truly suggested the shape of Jaguar's future, a lot of fans couldn't wait. Journalist Mel Nichols spoke for many in his report for Britain's *CAR* magazine: "Even without driving the Spyder, it was obvious that the one major element missing for so long from Brown's Lane, was back. Here was a sports car that, like the two-seater of old, simply screamed 'Jaguar!' even when standing still."

Encouraged, Pininfarina took the XJ Spyder home, made a "runner" out of what had been a non-functioning showpiece, and lightened its somewhat heavy appearance by covering the original dark "British Racing Green" paint with silver-grey. Nichols, among others, was invited down to Turin to drive it. Everybody apparently enjoyed the experience. One favorable observation was of very little air turbulence for an open-cockpit car. However, as the Spyder was just about as big as the standard XJ-S and weighed only about 150 pounds less, it was not a "real" sports car in the E-type mold.

Fortunately, it was not the real prototype F-type. Jaguar told Pininfarina thanks-but-no-thanks, and turned to a concept similar in nature but smaller and lighter. *CAR*, in a 1983 "spy" report, described this as using a shortened chassis taken from "XJ80," an allegedly imminent new BMW 5-Series-size sports sedan.

Such news may have been accurate then, but by June 1987, *AutoWeek* in Detroit was reporting that Coventry's comeback sports car would be built on the same assembly line as the new second-generation XJ-6. That sedan was Project XJ40; the new sports car was coded XJ41. Its engine would be the same new all-aluminum, twin-cam, 24-valve "Advanced Jaguar" inline six that would power XJ40, but possibly turbocharged. Later, a V-12 option might be added. Supplied with both hard and soft tops, the F-type body would initially be a two-seater,

*Pininfarina's one-off XJ Spyder of 1978 was built from a worn-out XJ-S. The Italian design house proposed that it fill the void left by retirement of the E-type in '74.*

*Jaguar's initial proposal for a new hard-core sports car was the "F-type." The subject of much rumor and spy photography during the mid and late '80s, it's still on hold.*

but with a 2+2 to follow. Again, however, all this suggested something larger and weightier than a "true" sports car. The press doodled up more sketches with shapes somewhere between the E-type's and the Pininfarina Spyder's, complete with the inevitable bullet nose. Production, said *AutoWeek*, might begin in 1990.

Britain's *Motor* then added four-wheel drive to the F-type stew. It would be the Ferguson Formula setup devised by Harry Ferguson and pioneered in production—albeit with scant commercial success—by the late-sixties Jensen FF. Jaguar was said to be testing the latest FF system in a specially modified XJ-S. *Motor* gave the F-type's introduction as "the early 1990s."

In October 1987, Jaguar's energetic new leader, John Egan, finally lent credence to the rumor-mongering by admitting to American trade weekly *Automotive News* that an "F-type" was definitely in the works, if still five years distant at best. "The componentry is coming along quite nicely," he explained, "but the problem is that we have a lot of things happening before that, which tend to crowd out a longer-term product each time."

Back in England the following February, *Performance Car* reported that said componentry included advanced composites. Jaguar was apparently concerned about weight now, so what the magazine called the "lightweight F-type" was unlikely ever to carry the V-12. This line of speculation seemed to be confirmed when *AutoWeek* weighed back in with news that because of excess pounds—both mass and money—"Jaguar top executives are having some second thoughts about their plans." The magazine said the F-type no longer was seen as a Porsche 928-type car but more a 944-fighter. Moreover, Jaguar might offer the car both as an all-out luxury four-wheel-drive and a simpler, slower, less expensive rear-driver. Neither would get the V-12, ever, but there would be a hot engine: a 4.0-liter, twin-turbo AJ6 tuned for a very lively 340 horsepower. What's more, said *AutoWeek*, the F-type would finally come to market in 1992.

Well, no, it wouldn't. Soon after Jaguar's $2.5-billion takeover by Ford in 1990, Dearborn decided the F-type as then envisioned would never work. There was just too much weight for the power—and too much money for the market.

Jaguar Chairman Bill Hayden, the long-time Ford Europe executive who replaced Egan at the beginning of July 1990, told *Autoweek* that, "The F-type was a beautiful-looking vehicle in terms of what it was, but the fundamental problem was the way it had been developed over a period of 10 years. During that time it got away from what the marketing department wanted and from what the engineers wanted. It was mid-way between a grand tourer and a sports car. We had aimed to do a sports car, not a grand tourer. Also we had lost a bit in terms of cost and the car was 400 pounds overweight. The F-type was getting close to the XJ-S in terms of cost structure, performance and weight." He went on to reveal that the car was originally seen as a 1986 or '87 model, that it had been

*In late '84 and early '85, plans were laid for a new-age Jaguar road rocket. Much wind tunnel work was needed if it were to reach its anticipated top speed, 220 mph.*

rescheduled for launch in 1993, and that it was now running at least three years tardier still. "[T]here was concern in the company that we had missed the opportunity we had been aiming for."

Hayden might have added that, although the F-type had been seen publicly only in "spy" photos of various prototypes, everyone knew it was based on sedan hardware then approaching the end of its useful commercial life. Besides, Jaguar's overseers surely included a faction arguing that the flagship role could be assumed by the spectacular XJ220 (more about the 220 in a moment).

Still, cancellation of the long-awaited F-type was taken as a bad sign by deep-dyed Coventry fans, who had not forgotten their suffering at Leyland's hands and feared just this kind of cost-cutting by the new American owners. But Karl Ludvigsen, a respected journalist, author, and auto-industry consultant quoted by *AutoWeek* in early 1991, took the view that Ford's action was courageous—a sign that it was serious about ensuring the future of its new subsidiary. "The F-type was too little too late," Ludvigsen observed, "an old car not up to the standards that have been set. Ford is interested in building a good Jaguar sports car, competitive in terms of weight, performance and cost."

So the idea is still alive, even if it has moved well down the "to do" list. There have been rumors that, as an interim measure, Coventry might let JaguarSport, its joint hot-car venture with Tom Walkinshaw, build a few examples of the "original" F-type after finishing a limited run of XJ220s. But it is generally understood that the wait for an F-type faithful to the generally understood meaning of "sports car" might now stretch into the next century.

In the meantime, there was work to be done on the bread-and-butter cars, and Jaguar's planned facelifts and refinements to existing models continued under Ford's auspices. The XJ-S was to be treated to a makeover—the second such in its long life—in the spring of 1991. The XJ-6, after an extensive engine-compartment redesign, was to offer V-12 power that autumn. Not so much for reasons of profitability as lack of the desired Jaguar aesthetics, the long-wheelbase "limo" concept was shelved pending an all-new platform.

That new platform, destined for a third-generation XJ-6-type sedan, is already under development as Project XJ90. Planned for introduction in 1995 or '96, it will compete in size, style, performance, and panache with BMW's 7-Series. Though XJ90 is said to retain much of the excellent XJ40 chassis, it will have a fresh floorpan and bodyshell. And power reportedly makes a complete break with roadgoing Jaguar tradition, involving a family of all-new V-type engines: a six certainly, with a derivative V-8 and perhaps a V-12, depending on market conditions and oil power politics.

V-engines make sense because they are shorter (and sometimes lighter) than inline powerplants, thus allowing more of a car's total length to be used for people and packages. But coming so soon after the Ford takeover, some Jaguar devotees feared it meant that these new "Jaguar" powerplants would be simply rebadged Dearborn or Ford Europe iron.

Indeed, their trepidation had a basis. In the interest of maximizing development funds, Ford once entertained the thought of bolting new Jaguar-designed cylinder heads atop some of its own well-tested blocks. But that scheme apparently failed to withstand the passionate defense of Jaguar chief engineer Jim Randle, a brook-no-compromises spirit of the same stripe as his tough old boss, Bob Knight, who for so long fought off British Leyland's clumsy attempts at "rationalization." Unlike Leyland, however, Ford went in knowing that Jaguar's continued worth, financial and otherwise, depended greatly on maintaining the make's clear, independent image—at least in production cars. On the racing front, a rules change did force Ford to hold the line on development costs. Thus, the Jaguar XJR-14, a sports-racer built for the 1991 season to suit a new international emphasis on non-turbocharged 3.5-liter engines, was powered by only a slightly modified version of Ford's well-proven Formula One V-8.

Actually, all automakers buy a large proportion of their components from outside suppliers. Jaguar had always done so, and as long as the primary parts were recognizable as Jaguar's, the cars remained Jaguars. Under the new owners, only such prosaic items as air-conditioning compressors, wiper and seat motors, door locks, and other things nobody cares much about—unless they're unreliable—would be sourced from Ford. And reliability, of course, now loomed paramount. The Brits had finally admitted they had work to do if they hoped to compete in this respect with the Germans and Japanese.

As for the XJ-S—that "golden oldie" in Sir John Egan's fond phrase—long-lived popularity had proven the basic type worth keeping. Thus an all-new XJ-S derived from XJ90 is scheduled for introduction about 18 months after the new sedan comes on-line.

And the "BMW fighter"? Speaking to *Automotive News* back in 1987, Egan had regretfully dismissed it. "That is really heavy metal. You need an awful lot of money to get into that . . . . We'd have to be very self-confident and very successful before we'd attempt to do anything in that area." Well, Ford infused both the money and the confidence. By 1991 it was known that a new sports-sedan challenger to the BMW 5-Series was in the works as another XJ90 spinoff, with introduction set for about 1998. Although it won't be quite as compact and racy as the revered old Mark II, it will fill a yawning gap in the range. While a name has yet to be decided, those involved are already said to be calling it "Mark III." In Jaguar's past, such monikers had a way of sticking.

"Product cycles," the shorter the better, have become a major competitive concern throughout an industry rocked by Japan's commitment to fresh models every four years or so. Generally, the revitalized Browns Lane aims to facelift its new models at four-year intervals

*The XJ220 was laid out much like a racing car, with a monocoque chassis of bonded aluminum incorporating a steel roll cage. The radiator would be in front, the fuel held in two rubber cells between cockpit and engine. Bodywork of hand-hewn aluminum would clothe it. "Jackknife" doors opening up and forward also were planned.*

and to replace them at age eight. This means relatively massive short- and long-term investment for a low-volume outfit. Which is precisely why Ford plans to recoup its spending by raising Jaguar's volume from 40,000 per year, the 1990 level, to 60,000 by the mid nineties and on to 150,000-200,000 by the millennium.

Until the first of the new production cats could be unleashed, something was needed to keep the press talking and the public dreaming about Jaguars. That job has fallen to two rare and rarefied limited editions destined to rank among the most memorable of Coventry's creations.

The very first of them, the XJ220, was disclosed in the fall of 1988 at the British International Motor Show. Describing its impact in *Car and Driver*, journalist Ray Hutton wrote that the XJ220 "captured the attention of the media and required unprecedented crowd-control measures at the Jaguar stand . . . . It made Jaguar the star of the Birmingham show."

The star was born nearly four years before, over the Christmas and New Year holidays of 1984-85. Chief engineer Jim Randle found himself mulling over the meaning to Jaguar of some international rally and race rules called Group B. The basic idea behind this was to reestablish something of the old commonality between competition and street cars. Group B was exciting technically, but the "Killer Bees" it fostered were quickly discredited as too fast for some kinds of competition—and many venues—and the class was soon abandoned. But two of the participating manufacturers redirected their development programs toward street-legal supercars. The results were spectacular corporate flag-wavers: Porsche's stupefyingly complex 959 and the stark, purposeful F40, the last Enzo Ferrari Ferrari.

Jaguar's head engineer itched to show his stuff, too. The nearest piece of paper happened to be a Christmas card. He turned it over and started sketching.

"The XJ220 began, in Randle's mind, as a type of machine that no longer exists: a roadgoing racer," wrote Hutton. "The kind of competition car you could drive to the circuit and then race, as the more adventurous chaps did in their D-types." But, of course, the car also would be a display piece for Jaguar's engineering expertise, and would, as Paul Frere noted in *Road & Track*, "project Jaguar tradition into the future." Randle took his sketches into work and wangled permission to turn them into metal, providing the job wouldn't interfere with anyone's normal schedule—and that it wouldn't cost anything.

Somehow, he managed it. Recalling the days of Sir William, Randle's enthusiasm electrified Coventry. A group of about a dozen specialists agreed to devote their off-hours; the "Saturday Club," they called themselves. Others throughout the sprawling factory found bits of hardware or some particular expertise to donate. Randle also made an evangelistic tour of the entire British auto industry. When it was all over, 46 different firms had contributed to the new supercar.

*As originally envisioned, the XJ220 was to have a longitudinally mounted 48-valve 6.2-liter V-12 derived from a racing engine and delivering about 520 bhp to all four wheels. Four-wheel steering and a suspension adjustable for ride height were included. Wheelbase was 112 inches and dry weight of the prototype was a portly 3448 pounds.*

Like the XK 120, XJ220 was named for its anticipated top speed. Randle laid it out as he would a racing car: a two-seat, mid-engine coupe with a monocoque chassis of bonded aluminum and incorporating a steel roll cage. He put the radiator in front, the fuel in two rubber-bag safety cells between cockpit and engine. Bodywork would be hand-hewn aluminum. Full-time four-wheel-drive was an initial part of the plan; so were four-wheel steering, anti-lock braking system, traction control, electronically adjustable shock absorbers, variable ride height, ground-effects venturis, automatically adjustable spoilers front and rear—all the high-tech tricks that automakers were yearning to try and learning to perfect in the mid eighties.

For power, Randle went to a storeroom for one of five special V-12s with quad-cam, 48-valve heads. The engines apparently were left over from a 1981-83 racing test program, though it is said they were actually far older, dating from the mid-sixties XJ13 experiments. Regardless, the engine selected happened to displace 6222 cc (380 cubic inches). After a moderate de-tune for street use, it made about 520 horsepower at 7000 rpm. The torque value was a little over 400 pounds/feet at 5000.

Mounted lengthwise, but not serving as a stressed chassis element as in the purpose-designed racers, the V-12 drove aft through a two-plate clutch to a five-speed transaxle. So far, that was like the competition XJR-series. But Randle got Ferguson (properly known as FF Developments), the four-wheel-drive experts already involved with Jaguar's other street-car programs, to run a long shaft through the vee of the engine and on forward beneath the center console to drive the front wheels. All four tires were the same size: 295/40ZR-17s on rims no less than eleven inches wide.

After more than three-and-a-half years of Saturdays and late nights, plus the last-minute scramble inherent to every such project, the XJ220 finally was ready for the Motor Show. It was worth the wait. What a sensational road bullet. No track testing had yet been done, but computer simulations projected 0-60-mph acceleration at 3.5 seconds, 0-100 at 8.0, and top speed of at least the magic 200 mph—and quite possibly the promised 220. At 211 mph, with the adjustable suspension lowered to its racing ride height of about three inches, the ground effects underbody would be generating a suction equal to the car's weight.

But unlike Ferrari's F40, this was no stripped supercar. The XJ220 boasted all the creature comforts a modern Jaguar buyer expected: air conditioning, power windows and mirrors, heated glass and seats, Connolly leather upholstery, high-quality audio, a tilt-adjustable steering wheel with five positions, even an infrared remote door-locking system. Those doors, by the way, were the exotic "jackknife" type familiar from the Lamborghini Countach, though Jaguar provided a dashboard switch to operate them. In short, the XJ220 was a marvel.

It also was a monster. Wheelbase sprawled to an incredible 112 inches—over 15 more than on the Countach. At 202 inches, its bodywork stretched more than *three feet* beyond the Lamborghini's 163. Width bulged to 78.7 inches (same as Countach's); height was 48.7 (versus 42.1). Good British beef, here; no wonder dry weight came to a meaty 3448 pounds. For all that, there was practically no room for luggage.

The price? Well, the car had been approved as a one-off demonstration of technical ability, and no firm commitment had been made to production. "But in the euphoria of launch day," reported Hutton in *Car and Driver*, "Egan upped the price every hour or so—along with his estimates of how many could be sold. The highest figures heard were $550,000 and a production run of 1000 cars."

Jim Randle was happy with the reception, too, though it was likely more than one observer might have said to him, "really, lovely car and all that, old boy, but honestly, in this age of oppressive speed limits just about everywhere, isn't a 200-mph supercar just the least bit absurd?"

"Yes, of course," was Randle's cheerful response, as quoted by Hutton. "If I was in the position of a customer with enough money to buy an XJ220, I would put it in a museum! I don't know where you could drive it properly, except on a track." It is believed that this remark did nothing to dampen the ardor of even a single checkbook-waving prospect.

But cold water soon doused the enthusiast fires anyway. It was not Ford who wielded the bucket; the wedding of Dearborn and Coventry still lay ahead. No, Jaguar itself took a grey-light-of-dawn look at the prototype XJ220, and saw it wasn't right. Not as a production model.

The main problem was that V-12 engine: beautiful, honorable, enjoyable, but very bulky. Tucked between the front wheels of an E-type, it had been fine; only the decision to provide 2+2 seating had run the Series 3 wheelbase up to 105 inches. The engine itself was no longer than the old XK six, so a 96-inch-wheelbase V-12E would have been entirely feasible. (And most desirable!) But mid-engine cars are hard to package so compactly, mainly because front wheelhouses tended to eat into leg room, and the snug-fitting XJ220 was a telling demonstration of what could go wrong. A physically shorter powertrain package was definitely needed.

Fortunately, there was one available. Looking forward to a future range of more efficient passenger cars, Jaguar had been working up a nice short, light V-6. The angle between its cylinder banks was 90 degrees—rather unusual for a V-6, where balance considerations make 60 degrees more common. Jaguar's comment was that the wider spread was necessary to limit overall height for installation beneath a low-profile hood in future sedans. The rest of the powerplant was impossible to fault: cast all in aluminum, with two camshafts per cylinder head and four valves per cylinder. That seemed to make it a born

*A second-generation XJ220 was announced in late 1989, this one with a twin-turbo 3.5-liter V-6 tuned to 500 bhp. This shorter engine helped reduce the wheelbase to 104 inches. Weight dropped to 2977 pounds. A switch to rear-drive, two-wheel steering, and a fixed-height suspension also was made. An altered side scoop shape helps identify the V-6 XJ220.*

racer. And in fact, it was under development in racing.

There were two displacements. For America's IMSA GTP series it was a 3.0-liter installed in an all-new car, the XJR-10. The 3.5-liter Group C version for Europe was very similar, but there were enough differences for the car it powered to be called XJR-11. (For Le Mans, the highly refined V-12 design continued in successful play aboard a model renamed the XJR-12.) In both cases the V-6 was turbocharged, and, though Tom Walkinshaw Racing was very guarded about specifics, either version was probably capable of something near 800 bhp. A spicy little powerplant indeed, and Randle's merry band of supercar chefs likely needed no urging at all to see what they could cook up with it.

This shorter engine prompted a second-generation XJ220 with seven big inches removed from wheelbase and eleven from overall length. That obviously saved on structural weight, and the lighter engine itself, plus a decision to delete four-wheel drive, further contributed to a substantial reduction in mass. As the streetgoing V-6 car could be turbocharged to the same 500-horse level as the previous 6.2-liter naturally aspirated V-12, the performance of the new XJ220 promised to be as much improved as its handling.

As announced by JaguarSport in December 1989, a production run of 220 cars was authorized at a price of £290,000 or upwards of $500,000 U.S., not including taxes. A left-hand-drive version would be available and the XJ220 would comply with motor vehicle regulations across Europe, but no orders would be accepted for the United States or Canada.

Structurally it was much as before, featuring what Jaguar described as a "bonded and riveted lightweight aerospace aluminum/aluminum honeycomb body/chassis" and aluminum exterior bodywork. In essence, the original aero package had survived, too, with wings at both ends (the front one tucked under the nose) and twin underbody venturi tunnels. However, Randle had reconsidered the need for so much downforce and had reduced the tunnel depth for less drag. Apparently, he also abandoned the adjustable ride-height feature. The revised car's drag coefficient was much better, from 0.38 or 0.35 depending on height setting, down to 0.32 with the new fixed ground clearance. An auxiliary benefit of the underbody reshaping was the chance to lower the exhaust system for increased luggage capacity. The rear-steering hardware had been deleted, too.

With bore and stroke of 94 × 84 mm (3.70 × 3.31 inches), the second-generation XJ220's V-6 displaced 3497.6 cc (213.4 cubic inches). Lubrication was by means of a racing-style dry-sump system, and there were other ultra-performance features, such as steel connecting rods and crankshaft. Thanks to twin water-cooled turbochargers feeding the engine through a pair of intercoolers, plus fuel injection and electronic engine management, horsepower was right at 500 at 6500 rpm, while the torque figure was 472 lbs/ft at 5000. That was with the standard catalytic exhaust converters, by the way, to suit the

*Anticipated top speed of the XJ220 is still above 200 mph, though the likely 0-60 mph time is 4.0 seconds, up from 3.5 for the V-12 version. With improved handling, more luggage space, and a load of luxury equipment among its charms, the XJ220 isn't likely to go begging for buyers willing to pay upwards of $800,000 for each of the 350 JaguarSport plans to produce.*

unleaded gasoline soon to be required across Europe, as in America. The prodigious power went through an 8.5-inch, twin-plate AP Racing clutch to a five-speed transaxle driving the rear wheels through a viscous-type limited-slip differential.

Wheelbase was down to a more manageable 104 inches, and overall length to 191. According to the official XJ220 spec sheet, height was reduced to 45 inches. Weight was now 23 pounds to the good side of 3000—somewhat short of the amazing reduction of more than 770 pounds indicated elsewhere in the publicity materials, but still a substantial improvement. That and deletion of the front-drive mechanism meant the rack-and-pinion steering could remain innocent of the power assistance contemplated for the 4×4 prototype. Further reducing steering effort were smaller front tires, though they were still pretty plump: Z-rated 245/40s on 9×17-inch rims. At the rear were huge 345/35s on wheels measuring 14 inches across and a whopping 18 in diameter.

Performance estimates still indicated a top speed above 200 and 0-100 acceleration of eight seconds flat. However, presumably because all that power had to flow through only the rear wheels now, the 0-60 projection rose half a second, to 4.0.

Jaguar discovered that abandoning the grand old V-12 did not turn off potential XJ220 owners, and it soon increased planned production. At this writing, the run will be 350 cars, all built by JaguarSport at an initial rate of two per week in a new factory at the town of Banbury. Having paid a deposit of $80,000, the first impatient customer expected to place hands on wheel in early 1992. Total price: "$800,000-plus," said *AutoWeek* late in 1990.

But, no, the XJ220 will still not be Federalized for North American sale.

Nor will the XJR-15. Just how did Jaguar come to fire *two* missilecars at the same time and at approximately the same target market? Before political smoke was blown over the scene, indications are that Tom Walkinshaw and his men took an independent look at the first V-12-powered XJ220 and determined to come up with something smaller, neater, and racier. After all, had they not built just such a vehicle and beaten the Group C racing world? And wasn't it time to encourage the wider commercial world to take a good look at TWR's expertise in the area of composite chassis construction?

Beginning in early 1988, Walkinshaw assigned a small team to alter an XJR racer into a V-12 road car. Reportedly, they started with the XJR-8LM that had been battered in a violent high-speed crash along the Mulsanne Straight in the 1987 Le Mans race. Rebuilt, rebodied and relabeled R9R, it served as a test mule for the concept. But two things soon became clear: It wasn't a very satisfactory concept for highway use; and to make it so was going to cost great gobs of money.

Work on the R9R stopped for a while as TWR took up the job of readying the XJ220 for production.

*Jaguar's new supercars boast racing blood. The XJ220's V-6 has ties to that of the XJR-11* (left) *and the XJR-10, shown at top in 1990 and far left in its '91 form. The XJR-15's V-12 has the single-cam heads proven in the '90 Le Mans-winning XJR-12* (above).

When that was put on temporary hold during 1989, Walkinshaw took another look at his own supercar, but now in light of an abandoned racing series called Procar.

Initiated by BMW, Procar basically was an inspired scheme to snatch corporate pride from the jaws of a potential competition disaster: the Bavarian firm's late-seventies M1. BMW had designed this mid-engine coupe for what was popularly known as "silhouette" racing, where one of the key homologation requirements was to build far more cars—400—than could ever be absorbed by the racing market alone. That dictated a design not only competitive enough for the track but practical enough for the street, with sufficient appeal to everyday, if particularly well-heeled, enthusiasts.

Unfortunately, the M1 project ran into long delays, and by the time the car was ready, the class was defunct. Fearing the image of its expensive supercar would sag and nobody would buy (though it was an excellent car), BMW formulated a one-model race series. Held in 1979 and 1980, Procar offered M1 owners a chance to compete in warmup races at selected Formula One Grand Prix events—against selected F1 drivers. The temptation proved compelling, the series got a lot of publicity, the racing itself turned out to be pretty good, and BMW salvaged the M1's reputation, if not its total investment.

As resurrected and revised by Major Tom and his mates at Browns Lane, the idea was called the JaguarSport Intercontinental Challenge. Here was the deal: JaguarSport would build a few—40 at first, later raised to 50—very high-performance, very finely crafted, very expensive cars suitable for both road and track. A further development of the R9R mule, they would be known as XJR-15s. To qualify to buy one, you had to agree to enter it in all three Intercontinental Challenge races. These would be supporting events for the season's three most heavily attended Grands Prix: Monaco, Silverstone, and Spa. Television coverage was assured. All race preparation and maintenance would be handled by TWR. The prize structure was such that the series-winning owner stood to earn back his or her original investment. Less tax and license, of course.

Anyone who knows racing people will recognize an irresistible proposition here, and will not be surprised to learn that, even before the first car was ready for sale, all 50 were spoken for. At one million U.S. dollars apiece. Plus tax and license, of course.

Really, it was a brilliant promotional scheme. First, it set JaguarSport's superexpensive supercar apart from the herd of a dozen or more rivals: a neo-Bugatti, a roadgoing McLaren, the V-16 Cizeta Moroder—even the XJ220. Amidst this rabble, the XJR-15 was something really special, descended from proven blood stock and able to demonstrate its prowess in battle.

From the customers' perspective, even those 49 who failed to get their million back could, after participating in the three mandatory races, drive happily off into the sunset. Or perhaps into safe storage; if ever there was a certain appreciating collectible it was the XJR-15.

First though, Walkinshaw actually had to transform the R9R into a reasonable roadgoing sports car. That turned out to be quite a job. Making it acceptable on the highway involved redesigning the rear cockpit structure to allow direct rearward vision not required on the track. The interior had to be wider and the door openings larger to make the wealthy buyers feel more welcome. The racer's carefully developed aerodynamics package had to go: That much downforce was counterproductive on the street, and the projecting rear wing was illegal in some countries anyway. R9R did follow both the racers and XJ220 in retaining twin underbody diffuser tunnels. But its entire body had to be raised to position lights and the added bumpers at legal heights and to allow enough ground clearance and wheel travel for a decent highway ride. Besides, some at Jaguar feared that TWR's ultra-scientific XJR-9 shape, which looked so good to the air, might strike potential customers as ugly. They called in a professional stylist. All in all, turning a Le Mans winner into a sports car was a lot harder than in the D-type/XK-SS days.

Doing it here called for a substantially revised tub and all-new body designs. Still, the clean-sheet XJR-15 owed more to its racing ancestors than did the XJ220. Constructed of carbon fiber and Kevlar, its chassis retained the V-12 engine as a stressed chassis member, and its transmission, suspension, steering, brakes, wheels and tires were all bona fide competition-quality hardware. The cockpit was pure racing business, too: a tight capsule of bare black-and-silver aerospace composites, hard form-fitting seats, utilitarian toggle switches, and a seriously anatomic little steering wheel whose lower rim was bent up for knee clearance.

The selected twelve displaced 5993 cc (365.7 cubic inches) and retained the race-proven single-cam, two-valve heads. The specification included a dry sump, forged-steel crankshaft and connecting rods, and forged aluminum pistons giving a compression ratio of 11.0:1. With racing-type electronic fuel injection and engine management system, nominal output was 450 bhp at 6250 rpm and 420 lbs/ft at 4500. There were three plates in the AP Carbon Racing clutch, and six straight-cut forward gears in the transaxle. Available gearing allowed a maximum velocity of up to 200 mph, but top speed in reality would be chosen to suit the specific circuit; a ceiling of 185 sounds more realistic given even the fastest of the three tracks to be used for the Challenge. Wheels measured 9.5×17-inches front, 13×17 rear, and would mount tires of "various Goodyear racing specifications," according to the official press release. Overall length was the 4.8 meters (188.9 inches) established by Group C regulations—Jaguar claimed the XJR-15 was technically eligible to race at Le

*An XJR-8LM wrecked at Le Mans and rebuilt as the R9R* (above) *was the test mule for the $1 million XJR-15* (opposite, top). *The XJR-15 is street-legal, but Jaguar will require buyers of all 50 built to run them in an exclusive road-race series. Very light and very fast, the XJR-14* (right) *has a 3.5-liter Formula One V-8 from Jaguar's new owner, Ford. This is the newest sports-racing standard-bearer for Coventry's proud cat.*

Mans—and width the 1.9 meters (74.8 inches) of the basic XJR-9 bodyshell. Height came to 43.3 inches, 2.7 taller than the real Le Mans car. Wheelbase was not stated in the Jaguar paperwork, but likely was the same 109.4 as the TWR racer's. Official weight was 1050 kilograms (2315 pounds). *Autocar & Motor* reported the prototype went 0-60 in 3.2 seconds, 0-150 in 13.5, and topped out at 190.

Wow. *Two* Hyper-Jags! How many determined trophy hunters could content themselves with just one? Performance to suit every taste: From lap-of-luxury XJ220 to lap-times-are-everything XJR-15, Jaguar had your mood swings covered.

Fine, but what about something ordinary enthusiasts could imagine themselves buying? Could either of these marvelous road monsters foreshadow a line of Jaguars for tomorrow?

One never knows. Chairman Egan once thanked Bob Tullius' Group 44 racing team for its efforts by stating, "The XJR-5 has done much to rekindle the spirit and rebuild the image of Jaguar at home and abroad, even though no production-model concept of the foreseeable future is likely to bear any relationship to it . . ." But only a few years later, JaguarSport was tooling up for short but real production runs of XJ220s and XJR-15s.

More significantly, the basic design concept of either could well lead to a true series-production, mid-engine Jaguar. If Browns Lane does start from scratch with another "F-type," it could do a lot worse than follow the lead of Honda's neat and nimble NSX, an ultra-modern machine that seems to have banished all the old mid-ships bugbears. Such a "Super Sports" done the Jaguar way with, say, a productionized version of the F1/Group C V-8, would continue the SS 100, XK 120, and E-type traditions in a way today's company could genuinely be proud of.

It was tradition that sparked the specially sport-tuned XJ-S and XJ-6 models introduced in 1990. Called the JaguarSport 6.0-liter XJR-S Coupe and 4.0-liter XJR Saloon respectively, each offered more than cosmetic performance enhancements: stronger engines, tauter suspensions, sharper steering, wider tires—altogether a more muscular presence on the road. Enthusiast magazines were soon saying that these demonstrated the way all Jaguars should go, handle, and look.

Given luck, and spirit, and the continued fierce loyalty of the sort displayed so often over so many years, they will. It is just that combination of elements that defines the aura of inevitability about Jaguar, the feeling that this marque was not created, but sprang spontaneously into being to satisfy a natural need. After so much trial, having survived so much adversity, the proud leaping cat of Coventry, once endangered but now stronger than ever in many ways, remains a unique and precious part of the automotive realm. Long may it prowl.

# Specifications of Selected Jaguars

## JAGUAR ENGINES

| | |
|---|---|
| **Type** | **XK-6** |
| Configuration | I-6, iron block, aluminum head with 2 overhead camshafts, 2 valves per cylinder |

| | | | | | |
|---|---|---|---|---|---|
| Size categories (liters) | 2.4 | 2.8 | 3.4 | 3.8 | 4.2 |
| Displacements (cc) | 2483 | 2792 | 3442 | 3781 | 4235 |
| Bores (mm) | 83 | 83 | 83 | 87 | 92.07 |
| Strokes (mm) | 76.5 | 86 | 106 | 106 | 106 |
| Best bhp (pass. cars) | 120 | 180 | 250 | 265 | 265 |
| @ rpm | 5750 | 6000 | 5500 | 5500 | 5400 |

| | |
|---|---|
| **Type** | **V-12** |
| Configuration | 60-deg. V-12, aluminum block and heads, 1 overhead camshaft per head, 2 valves per cylinder |

| | | |
|---|---|---|
| Size categories (liters) | 5.3 | 6.0 (plus racing versions to 7.0) |
| Displacments (cc) | 5343 | 5993 |
| Bores (mm) | 90 | 90 |
| Strokes (mm) | 70 | 78.5 |
| Best bhp (pass. cars) | 295 | 318 (DIN) |
| @ rpm | 5500 | 5250 |

| | |
|---|---|
| **Type** | **AJ-6** |
| Configuration | I-6, aluminum block and head with 2 overhead camshafts, 4 valves per cylinder |

| | | | | |
|---|---|---|---|---|
| Size categories (liters) | 2.9 (1cam) | 3.2 | 3.6 | 4.0 |
| Displacements (cc) | 2919 | 3239 | 3590 | 3980 |
| Bores (mm) | 91 | 91 | 91 | 91 |
| Strokes (mm) | 74.8 | 83 | 92 | 102 |
| Best bhp | 165 | 200 (DIN) | 221 | 248* |
| @ rpm | 5600 | 5250 | 5000 | 5150 |

*(JaguarSport model)

| | |
|---|---|
| **Type** | **V-6** |
| Configuration | 90-deg. V-6, aluminum block and heads with 2 overhead camshafts per head, 4 valves per cylinder (plus 2 turbochargers in all applications to date) |
| Size categories (liters) | 3.5 (3.0 racing version also exists) |
| Displacement (cc) | 3498 |
| Bore (mm) | 94 |
| Stroke (mm) | 84 |
| Best bhp (XJ220) | 500 |
| @ rpm | 6500 |

| | |
|---|---|
| **Model name** | **SS Jaguar 100** |
| Years produced | 1935-39 |
| Number built | 308 |
| Configuration | Front engine; 2-seat-plus-shelf. 1 body style: roadster |
| Engine | Standard-based I-6, 2.7 or 3.5 with pushrod OHV, max 125 bhp |
| Suspension, front | Beam axle with leaf springs |
| rear | Live axle with leaf springs |
| Wheelbase (in.) | 104 |
| Track, front (in.) | 54 |
| rear (in.) | 54 |
| Length (in.) | 153 |
| Best 0-60 mph (sec.) | 10.4 |
| Best top speed (mph) | 101 |

| | |
|---|---|
| **Model name** | **XK 140** |
| Years produced | 1954-57 |
| Number built | 8,884 |
| Configuration | Front engine; 2-seat. 3 body styles: roadster, coupe, convertible |
| Engine | XK-6 3.4 |
| Suspension, front | Independent with torsion bars |
| rear | Live axle with leaf springs |
| Wheelbase (in.) | 102 |
| Track, front (in.) | 51.5 |
| rear (in.) | 51.4 |
| Length (in.) | 176 |
| Best 0-60 mph (sec.) | 8.4 |
| Best top speed (mph) | 129 |

| | |
|---|---|
| **Model name** | **XK 120** |
| Years produced | 1949-54 |
| Number built | 12,078 |
| Configuration | Front engine; 2-seat. 3 body styles: roadster, coupe, convertible |
| Engine | XK-6 3.4 |
| Suspension, front | Independent with torsion bars |
| rear | Live axle with leaf springs |
| Wheelbase (in.) | 102 |
| Track, front (in.) | 51 |
| rear (in.) | 50 |
| Length (in.) | 173 |
| Best 0-60 mph (sec.) | 8.5 |
| Best top speed (mph) | 125 |

| | |
|---|---|
| **Model name** | **XK 150, XK 150S** |
| Years produced | 1957-60 |
| Number built | 9395 |
| Configuration | Front engine; 2-seat and 2+2. 3 body styles: roadster, coupe, convertible |
| Engine | XK-6 3.4; XK 150S, 3.8 |
| Suspension, front | Independent with torsion bars |
| rear | Live axle with leaf springs |
| Wheelbase (in.) | 102 |
| Track, front (in.) | 51.25 |
| rear (in.) | 51.25 |
| Length (in.) | 177 |
| Best 0-60 mph (sec.) | 7.3 |
| Best top speed (mph) | 136 |

(All models 2-door unless noted; all models rear-drive; production numbers per A. Whyte or P. Porter; performance figures per various published tests)

| | |
|---|---|
| **Model name** | **XK-SS** |
| Years produced | 1957 |
| Number built | 16 |
| Configuration | Front engine; 2-seat roadster |
| Engine | XK-6 3.4 |
| Suspension, front | Independent with torsion bars |
| rear | Live axle with torsion bar |
| Wheelbase (in.) | 90 |
| Track, front (in.) | 50 |
| rear (in.) | 48 |
| Length (in.) | 166 |
| Best 0-60 mph (sec.) | 5.2 |
| Best top speed (mph) | 149 |

| | |
|---|---|
| **Model name** | **E-type Series 2** |
| Years produced | 1969-70 |
| Number built | 18,820 |
| Configuration | Front engine; 2-seat and 2+2. 3 body styles: coupe, convertible, 2+2 coupe |
| Engine | XK-6 4.2 |
| Suspension, front | Independent with torsion bars |
| rear | Independent with 4 coil springs |
| Wheelbase (in.) | 96 (105 for 2+2) |
| Track, front (in.) | 50 |
| rear (in.) | 50 |
| Length (in.) | 175.5 (184 for 2+2) |
| Best 0-60 mph (sec.) | 7.0 |
| Best top speed (mph) | 150 |

| | |
|---|---|
| **Model name** | **E-type (XK-E in U.S.) Series 1** |
| Years produced | 1961-68 |
| Number built | 38,410 |
| Configuration | Front engine; 2-seat and 2+2. 3 body styles: coupe, convertible, 2+2 coupe (2+2 from '65, 4.2 only) |
| Engine | XK-6 3.8 and 4.2 |
| Suspension, front | Independent with torsion bars |
| rear | Independent with 4 coil springs |
| Wheelbase (in.) | 96 (105 for 2+2) |
| Track, front (in.) | 50 |
| rear (in.) | 50 |
| Length (in.) | 175.5 (184 for 2+2) |
| Best 0-60 mph (sec.) | 6.9 |
| Best top speed (mph) | 153 |

| | |
|---|---|
| **Model name** | **E-type Series 3** |
| Years produced | 1971-75 |
| Number built | 15,290 |
| Configuration | Front engine; 2-seat and 2+2. 2 body styles: convertible, 2+2 coupe |
| Engine | V-12 5.3 |
| Suspension, front | Independent with torsion bars |
| rear | Independent with 4 coil springs |
| Wheelbase (in.) | 105 |
| Track, front (in.) | 54.5 |
| rear (in.) | 53 |
| Length (in.) | 184 |
| Best 0-60 mph (sec.) | 6.4 |
| Best top speed (mph) | 146 |

## SPORTS CARS (cont.)

| | |
|---|---|
| **Model name** | **XJ-S** |
| Years produced | 1975- |
| Number built | Currently in production |
| Configuration | Front engine; 2+2 and 2-seat. 3 body styles: 2+2 coupe, 2-seat cabriolet, 2-seat convertible |
| Engine | V-12 5.3 and 6.0 (JaguarSport); AJ-6 3.6 |
| Suspension, front | Independent with coil springs |
| rear | Independent with 4 coil springs |
| Wheelbase (in.) | 102 |
| Track, front (in.) | 58 |
| rear (in.) | 58.5 |
| Length (in.) | 191.5 |
| Best 0-60 mph (sec.) | 6.5 (JaguarSport) |
| Best top speed (mph) | 158 (JaguarSport) |

| | |
|---|---|
| **Model name** | **XJ220** (production version; built by JaguarSport, England) |
| Years produced | (to begin production in 1992) |
| Number built | (350 scheduled) |
| Configuration | Mid-engine; 2-seat coupe |
| Engine | V-6 3.5 turbocharged |
| Suspension, front | Independent with coil springs |
| rear | Independent with coil springs |
| Wheelbase (in.) | 104 |
| Track, front (in.) | NA |
| rear (in.) | NA |
| Length (in.) | 191 |
| Best 0-60 mph (sec.) | 4.0 (projected) |
| Best top speed (mph) | 200-plus (projected) |

## SEDANS

| | |
|---|---|
| **Model name** | **SS1** |
| Years produced | 1932-36 |
| Number built | 4,230 |
| Configuration | Front engine; 2+2 and 4-seat. 5 body styles: coupe, convertible, open tourer, sedan, "Airline" sedan |
| Engine | Standard-based I-6 2.1 to 2.7 with L-head, max 75 bhp |
| Suspension, front | Beam axle with leaf springs |
| rear | Live axle with leaf springs |
| Wheelbase (in.) | 112 to 119 (note: 3 chassis versions through model run) |
| Track, front (in.) | 49 to 53 |
| rear (in.) | 49 to 53 |
| Length (in.) | NA |
| Best 0-60 mph (sec.) | NA |
| Best top speed (mph) | 82 |

(note: smaller SS 2 range also available)

| | |
|---|---|
| **Model name** | **SS Jaguar Saloon** |
| Years produced | 1936-40, 1946-49 (postwar: "Mark IV") |
| Number built | 13,268 |
| Configuration | Front engine; 4-seat. 2 body styles: 4-door sedan, 2-door convertible |
| Engine | Standard-based I-6 2.7; 3.5 as SS 100, max 125 bhp |
| Suspension, front | Beam axle with leaf springs |
| rear | Live axle with leaf springs |
| Wheelbase (in.) | 119 and 120 (note: 2 chassis/body constructions through model run) |
| Track, front (in.) | 54 and 56 |
| rear (in.) | 54 and 56 |
| Length (in.) | 173 and 186 |
| Best 0-60 mph (sec.) | 16.8 |
| Best top speed (mph) | 91 |

(note: "1½ litre" range also available)

| | |
|---|---|
| **Model name** | **Mark V** |
| Years produced | 1949-51 |
| Number built | 10,466 |
| Configuration | Front engine; 4-seat. 2 body styles: 4-door sedan, 2-door convertible |
| Engine | Standard-based I-6 2.7 and 3.5 pushrod OHV, max 125 bhp |
| Suspension, front | Independent with torsion bars |
| rear | Live axle with leaf springs |
| Wheelbase (in.) | 120 |
| Track, front (in.) | 56.5 |
| rear (in.) | 57.5 |
| Length (in.) | 187 |
| Best 0-60 mph (sec.) | 14.7 |
| Best top speed (mph) | 92 |

| | |
|---|---|
| **Model name** | **Mark X; 420G** |
| Years produced | 1962-66; 1967-70 |
| Number built | 24,282 |
| Configuration | Front engine; 4-door, 5-seat sedan; limousine also available |
| Engine | XK-6 3.8, 4.2 |
| Suspension, front | Independent with torsion bars |
| rear | Independent with 4 coil springs |
| Wheelbase (in.) | 120 |
| Track, front (in.) | 58 |
| rear (in.) | 58 |
| Length (in.) | 202 |
| Best 0-60 mph (sec.) | 10.4 |
| Best top speed (mph) | 123 |

| | |
|---|---|
| **Model name** | **Mark VII and VIIM; VIII; IX** |
| Years produced | 1951-57; 1957-59; 1959-61 |
| Number built | 30,969; 6,212; 10,009 |
| Configuration | Front engine; 4-door, 5-seat sedan |
| Engine | XK-6 3.4, 3.4, 3.8 |
| Suspension, front | Independent with torsion bars |
| rear | Live axle with leaf springs |
| Wheelbase (in.) | 120 |
| Track, front (in.) | 56 to 58 |
| rear (in.) | 56.5 to 58 |
| Length (in.) | 196.5 |
| Best 0-60 mph (sec.) | 11.3 |
| Best top speed (mph) | 114 |

| | |
|---|---|
| **Model name** | **Daimler Limousine** |
| Years produced | 1968- |
| Number built | 4642 through 1990 |
| Configuration | Front engine; 4-door limousine |
| Engine | XK-6 4.2 |
| Suspension, front | Independent with torsion bars |
| rear | Independent with 4 coil springs |
| Wheelbase (in.) | NA |
| Track, front (in.) | NA |
| rear (in.) | NA |
| Length (in.) | NA |
| Best 0-60 mph (sec.) | NA |
| Best top speed (mph) | NA |

## SEDANS (cont.)

| | |
|---|---|
| **Model name** | **Mark II** (includes Mark I, S-type, 240, 340, 420, plus Daimler versions) |
| Years produced | 1956-69 |
| Number built | 185,200 |
| Configuration | Front engine; 4-door, 5-seat sedan |
| Engine | XK-6 2.4, 3.4, 3.8, 4.2; 2.5 Daimler V-8 with pushrod OHV |
| Suspension, front | Independent with coil springs |
| rear | Live axle on quarter-elliptic leaf springs; later independent with 4 coils |
| Wheelbase (in.) | 107 to 108 |
| Track, front (in.) | 54.6 to 55 |
| rear (in.) | 50.6 to 55 |
| Length (in.) | Various |
| Best 0-60 mph (sec.) | 8.5 |
| Best top speed (mph) | 125 |

| | |
|---|---|
| **Model name** | **XJ6 and XJ12,** including Coupe models of both (Daimler versions also available) |
| Years produced | 1969-86 (V-12 still in production) |
| Number built | Over 320,000, including 47,400 V-12 through 1990 |
| Configuration | Front engine; 4-door, 5-seat sedan (3 Series offered through model run) |
| Engine | XK-6 2.8, 3.4, 4.2; V-12 5.3 |
| Suspension, front | Independent with coil springs |
| rear | Independent with 4 coils |
| Wheelbase (in.) | 108.8 to 112.8 |
| Track, front (in.) | 58 |
| rear (in.) | 58.3 |
| Length (in.) | 189.6 to 195.2 |
| Best 0-60 mph (sec.) | 8.6 |
| Best top speed (mph) | 148 |

| | |
|---|---|
| **Model name** | **XJ6 (XJ40)** including JaguarSport model, plus Daimler versions |
| Years produced | 1986- |
| Number built | Currently in production |
| Configuration | Front engine; 4-door, 5-seat sedan |
| Engine | AJ-6 2.9, 3.6, 4.0 |
| Suspension, front | Independent with coil springs |
| rear | Independent with 2 coils |
| Wheelbase (in.) | 113 |
| Track, front (in.) | 59.1 |
| rear (in.) | 59 |
| Length (in.) | 196.4 |
| Best 0-60 mph (sec.) | 7.2 (Mfg. est.) |
| Best top speed (mph) | 145 (Mfg. est.) |

## RACING CARS

| | |
|---|---|
| **Model name** | **C-type** |
| Years produced | 1951-53 |
| Number built | 54 |
| Configuration | Front engine; 1-door, 2-seat roadster |
| Engine | XK-6 3.4 |
| Suspension, front | Independent with torsion bars |
| rear | Live axle with torsion bar |
| Wheelbase (in.) | 96 |
| Track, front (in.) | 51 |
| rear (in.) | 51 |
| Length (in.) | 157 |
| Best 0-60 mph (sec.) | 8.0 |
| Best top speed (mph) | 149 |

| | |
|---|---|
| **Model name** | **D-type** |
| Years produced | 1954-56 |
| Number built | 71 (not including 16 XK-SS) |
| Configuration | Front engine; 1-door, 2-seat roadster |
| Engine | XK-6 3.0, 3.4, 3.8 |
| Suspension, front | Independent with torsion bars |
| rear | Live axle with torsion bar |
| Wheelbase (in.) | 90 |
| Track, front (in.) | 50 |
| rear (in.) | 48 |
| Length (in.) | 154 to 161.5 |
| Best 0-60 mph (sec.) | 4.7 |
| Best top speed (mph) | 179 |

| | |
|---|---|
| **Model name** | **XJ13** |
| Years produced | circa 1966 |
| Number built | 1 |
| Configuration | Mid-engine; 2-seat roadster |
| Engine | DOHC V-12 5.0-liter, 502 bhp |
| Suspension, front | Independent with coil springs |
| rear | Independent with coil springs |
| Wheelbase (in.) | 96 |
| Track, front (in.) | 56 |
| rear (in.) | 56 |
| Length (in.) | 176.5 |
| Best 0-60 mph (sec.) | NA |
| Best top speed (mph) | 175 |

| | |
|---|---|
| **Model name** | **Lightweight E-type** |
| Years produced | 1963-64 |
| Number built | 12 approx. (plus 1 E2A unveiled 1960) |
| Configuration | Front engine; 2-seat hardtop, also coupe |
| Engine | XK-6 3.8 with aluminum block (E2A, 3.0) |
| Suspension, front | Independent with torsion bars |
| rear | Independent with 4 coil springs |
| Wheelbase (in.) | 96 |
| Track, front (in.) | NA (varied with tires) |
| rear (in.) | NA (varied with tires) |
| Length (in.) | NA (varied with bodywork) |
| Best 0-60 mph (sec.) | NA |
| Best top speed (mph) | 176 |

| | |
|---|---|
| **Model name** | **XJR-5; XJR-7** (built by Group 44, USA) |
| Years produced | 1982-85; 1985-87 |
| Number built | 11; 5 |
| Configuration | Mid-engine; 2-seat coupe |
| Engine | V-12, various displacements 5.5 to 7.0 |
| Suspension, front | Independent with coil springs |
| rear | Independent with coil springs |
| Wheelbase (in.) | 108.5 |
| Track, front (in.) | 65 |
| rear (in.) | 3 |
| Length (in.) | 186 |
| Best 0-60 mph (sec.) | NA |
| Best top speed (mph) | 219 (est.) |

| | |
|---|---|
| **Model name** | **XJR-6, XJR-8 and -8LM; XJR-9 and -9LM; XJR-12** (built by TWR, England) |
| Years produced | 1985 |
| Number built | NA |
| Configuration | Mid-engine; 2-seat coupe |
| Engine | V-12 various displacements 6.0 to 7.0 |
| Suspension, front | Independent with coil springs |
| rear | Independent with coil springs |
| Wheelbase (in.) | 109.4 |
| Track, front (in.) | NA (varied with tires and model) |
| rear (in.) | NA (varied with tires and model) |
| Length (in.) | 188.9 |
| Best 0-60 mph (sec.) | NA |
| Best top speed (mph) | 244 |

| | |
|---|---|
| **Model name** | **XJR-15** (built by JaguarSport, England) |
| Years produced | 1991 |
| Number built | 50 (scheduled) |
| Configuration | Mid-engine; 2-seat coupe |
| Engine | V-12 6.2 |
| Suspension, front | Independent with coil springs |
| rear | Independent with coil springs |
| Wheelbase (in.) | 109.4 |
| Track, front (in.) | NA |
| rear (in.) | NA |
| Length (in.) | 188.9 |
| Best 0-60 mph (sec.) | 3.2 (est., per test of prototype) |
| Best top speed (mph) | 185-200 (established by gearing) |

| | |
|---|---|
| **Model name** | **XJR-10 and XJR-11** (built by TWR, England) |
| Years produced | 1989 |
| Number built | NA |
| Configuration | Mid-engine; 2-seat coupe |
| Engine | V-6 3.0 and 3.5 turbocharged |
| Suspension, front | Independent with coil springs |
| rear | Independent with coil springs |
| Wheelbase (in.) | 109.4 |
| Track, front (in.) | NA |
| rear (in.) | NA |
| Length (in.) | 188.9 |
| Best 0-60 mph (sec.) | NA |
| Best top speed (mph) | NA (established by gearing) |

| | |
|---|---|
| **Model name** | **XJR-14** |
| Years produced | 1991 |
| Number built | NA |
| Configuration | Mid-engine; 2-seat coupe |
| Engine | DOHC V-8 3.5, based on Ford F1 |
| Suspension, front | Independent with torsion bars |
| rear | Independent with coil springs |
| Wheelbase (in.) | NA |
| Track, front (in.) | NA |
| rear (in.) | NA |
| Length (in.) | 188.9 |
| Best 0-60 mph (sec.) | NA |
| Best top speed (mph) | NA |

# INDEX

£6.99

Published by Alligator Books Ltd
Gadd House, Arcadia Avenue
London N3 2JU
Printed in China

SPIDER-MAN 2™

# Contents

## Peter Parker/Spider-Man

Peter Parker was brought up in Queens by his Aunt May and Uncle Ben. Peter was a shy, timid student. That was until he was bitten by a genetically-altered spider during a school trip. Peter was never the same again. Mysteriously, the spider bite endowed the young student with extraordinary powers. Peter was possessed with extreme strength and the agility of a spider.

Spider-Man was born.

At first, Peter took advantage of his powers to make money. But the tragic death of Uncle Ben made Peter realize the importance of using his unique strength and skills to fight crime. He tried hard to heed the wise words of his uncle: "With great power comes great responsibility."

Spider-Man climbs walls, leaps across roof-tops and senses danger from afar. He can defeat the most dangerous criminals and protect the innocent from harm. But for college student Peter, being a hero is not so easy.

Peter's secret life as Spider-Man leaves him little time to concentrate on his studies. And he has little time for the people he loves. Peter's relationships with those closest to him are falling apart. He longs to tell Mary Jane, the girl of his dreams, about his dual identity, but he dares not. It could put her life in danger. Peter's friendship with Harry Osborn is strained by Harry's bitterness over his father's death and his gnawing hatred towards Spider-Man. Even Peter's relationship with his beloved Aunt May is starting to show cracks.

Now Peter faces one of the toughest challenges of his life. He comes face to face with the demonic Doc Ock. Can Spider-Man defeat the multi-tentacled menace? Will Peter be able to balance life as a student with life as Spider-Man? And will he declare his love for Mary Jane?

# DR. OTTO OCTAVIUS/DOC OCK

Dr. Otto Octavius is a brilliant scientist. He is a charming, gentle man, brimming with vitality and enthusiasm. Among fellow scientists, he has a reputation as a hard-working, honest man with a brilliant mind.

Dr. Octavius' latest project is dedicated to experiments using fusion as a new source of energy. Peter Parker is writing his college science paper on the scientist's theories and experiments.

And now Dr Octavius is on the brink of unveiling a major scientific breakthrough. The first demonstration of his mastery is to be held in the laboratory that he and his wife Rosie have painstakingly built up over many years.

But the super scientist's dreams are about to come crashing down. His demonstration goes disastrously wrong, and with the most terrible consequences. The gentle genius is transformed into Doc Ock, a multi-tentacled terror.

The maniacal Doc Ock is determined to salvage his scientific reputation. To what lengths will the doctor go to achieve his deranged dreams?

# MARY JANE WATSON

Throughout her childhood, Mary Jane Watson (MJ to her friends) lived next door to Peter. Behind a happy smile, Mary Jane hid the truth about her troubled home life and her abusive father.

Mary Jane always felt able to trust and confide in her friendly, gentle neighbour, Peter Parker. But the feisty red-head did not know that, for a long time, Peter had secretly been in love with her.

When some villains threatened the aspiring actress, Spider-Man swung to her rescue. For a while, Mary Jane thought she was in love with the webbed wonder. Soon she realized that her feelings were in fact for her childhood friend and neighbour, Peter. But Peter knew that association with him could put her life in danger, and a single kiss between them led nowhere.

Now Mary Jane lives in New York where she is working hard to achieve her dream of being a successful actress.

Mary Jane still sees Peter but he has let her down too many times. Can she and Peter remain friends?

# HARRY OSBORN

Harry Osborn and Peter Parker are best friends.

Like all friends, they have had their ups and downs. However hard Harry tried, he could never live up to his father's expectations. Harry's father, Norman Osborn, always seemed more impressed by Harry's buddy, diligent student Peter Parker. Like Peter, Harry has always held strong feelings for Mary Jane and for a while Mary Jane was Harry's girlfriend. Despite these difficulties, Harry and Peter remained close buddies.

The appearance of Spider-Man put a much greater strain on their friendship. Spider-Man had defeated the devilish Green Goblin, a terrible persona that had overtaken Harry's father. Since then, Harry has harboured a hatred for Spider-Man, blaming him for the death of his father.

Today, Harry is in charge of special projects at OsCorp, the company once headed by his father. OsCorp is financing Dr. Octavius' latest research and it is Harry who introduces his friend Peter to the great scientist.

Can Harry and Peter's friendship survive Harry's festering hatred of Peter's alter ego?

## AUNT MAY

For many long years, Aunt May was happily married to Ben. She and her kindly husband brought up Peter Parker, their orphaned nephew. Ben was killed in tragic circumstances, gunned down by a crook that Peter had allowed to escape.

The death of Ben turned Aunt May's life upside-down. She struggled financially to meet all the bills. And sadly she found that her once devoted nephew, Peter Parker, didn't seem to have any time for her.

Will Aunt May be able to cope?

## J JONAH JAMESON

J Jonah Jameson is the moody editor at the tabloid newspaper, the *Daily Bugle*.

Peter Parker works for him as a photographer and to Jameson's amazement Parker is the only one who manages to get great action pics of Spider-Man.

Jameson constantly pressurizes Peter to take more and better photographs of Spider-Man. Jameson uses his newspaper to question Spider-Man's motives.

He has no qualms about portraying the Spider-Man as a dangerous vigilante, who creates more crime in the city.

# Caught in a Web

**Some words are caught in Spider-Man's web.** Can you work out which words are hidden in the tangle? You'll find them written horizontally, vertically, diagonally and even backwards.

Use a pencil to circle or draw a line through the words you find.

**The words to look for are:**

Web
Spinner
Power
Responsible
Wall
Climber
Danger
Hero
Crawler

| | | | | | | | | | | |
|---|---|---|---|---|---|---|---|---|---|---|
| B | D | U | L | L | A | W | U | E | C | D |
| C | M | T | A | V | P | G | P | I | R | N |
| L | T | F | J | X | D | K | C | W | A | L |
| H | U | R | M | O | A | Y | L | E | W | I |
| R | E | S | P | O | N | S | I | B | L | E |
| S | N | R | T | U | G | A | M | L | E | G |
| R | I | D | O | N | E | O | B | B | R | C |
| Q | U | Y | L | C | R | J | E | V | S | V |
| B | P | T | P | O | W | E | R | E | U | Z |
| U | W | F | K | R | E | N | N | I | P | S |



Answers: page 60

# Make a Match!

There are 12 silhouettes. Can you match up the identical ones?

Answers: page 60

# SPIDER-MAN 2™
## The Story

Based on the Motion Picture
Screenplay by Alvin Sargent
Screen Story by Alfred Gough & Miles Millar and Michael Chabon
Based on the Marvel Comic Book by Stan Lee and Steve Ditko

## CHAPTER 1

Two years ago, life changed dramatically for mild-mannered Peter Parker when, in a twist of fate, a spider's bite gave him phenomenal powers. The timid college student was suddenly able to spin webs, climb walls and sense danger, just like a spider. He was possessed with super human strength.
Spider-Man was born.

Life was never the same again for meek Peter Parker. And it was not always easy. He had to learn how to use these unique powers wisely.

Now, Peter struggles to balance his dual identities as regular guy Peter Parker and Spider-Man, super web slinging spider hero who resolutely protects his fellow men and women from crime.

Life is complicated enough, even for everyday folk. But for Peter, there are not enough hours in the day. He has to be a good friend, a loyal nephew and a diligent student as well as literally jumping to the defense of innocent city dwellers and fighting off crooks, villains and diabolic maniacs.

For Peter, it was just another average Monday morning. He was on his way to the offices of the Bugle, the newspaper where he sometimes worked as a photographer to earn some money.

Peter took a shortcut on his scooter down a dark deserted alley. But suddenly he could feel his body tingle. His spidey senses were on alert. There was trouble somewhere around. He couldn't yet work out what was up.

As Peter drove down the street, he heard the excited screams of children playing. It made a welcome change from the screams of terror that were heard too often in the city. It reminded Peter of the times when he and his neighbour Mary Jane played together as children. How simple and uncomplicated life had been then!

"Chase!"

"Gotcha!"

"Your turn."

"Catch."

Peter could hear the joyful shouts more easily as the children chased each other round the corner. They were so engrossed in their game that they took no notice of the cars and vans in the busy city streets.

Peter's extra sensitive hearing was on high alert. He could hear a thundering rumbling noise. It was getting closer. He anxiously turned around, sensing danger.

The children's ball rolled into the main road. Without a thought, the two children ran straight out into the road to retrieve their brightly coloured ball.

They hadn't seen or heard the huge juggernaut charging down the road. The driver of the massive truck had spotted the ball and children, but it was too late. He would not be able to brake in time. The children suddenly looked up in terror at the storming advance of the oncoming truck.

Instantly, a giant web looped through the air. It scooped down and hauled the stunned children to safety.

Drivers screeched to a halt. Horns blared. Passers-by stared in.

Spider-Man had done it again.

The driver of the juggernaut put his head in his hands and silently thanked Spider-Man.

"Kids, the park's a better and safer place to play," suggested Spider-Man as he untangled the shocked children from the sticky web. And then he was off. "Sure thing Spider-Man, and thanks for saving us," called the kids. "Hey, mister," they shouted after him, "that was some cool ride in your web!"

Drat. Peter was going to be late. He jumped onto his scooter and resolved to go straight to the offices of the Bugle where he worked. Peter pulled up at the red traffic lights, impatient for them to turn green so he could get going. But before long, Peter's spidey senses were on red alert for the second time that Monday morning. Peter looked around.

A small elderly man was standing at a cash machine waiting for the machine to dispense his money. Two thugs were hovering nearby the frail gentleman. As the machine ejected the money, the crooks violently pushed the man to the ground and snatched the notes.

THWACK! BANG! KERPLUNK!

In seconds, Spider-Man made mincemeat of the vicious villains. He helped the elderly man to his feet and returned his money to him. "Thank you" stuttered the bewildered man.

The police arrived, car wheels screeching, sirens blaring. Protesting and struggling, the low-lifes were taken away to face justice. Spider-Man swung high and low, and then back to life as plain old Peter Parker.

Peter jumped back on his scooter. Blast! The traffic lights were on red again.

Peter entered the office of the Bugle newspaper. The editor, the gruff Jonah Jameson was always asking Peter for more and better pictures of Spider-Man.

No other photographer seemed able to get such amazing action-packed Spidey photos. But Peter was getting fed up of it. He wanted to show a different side to his city, New York.

"Let's show something else", suggested Peter to the grumpy Jameson. "What about some photographs of kids playing? You see, Spider-Man thinks that all of these pictures are turning New Yorkers against him."

"Who cares?" bellowed Jameson, practically blowing poor Peter out of his office. "Spider-Man pictures sell, and that's what people want."

There was no budging the man. "Get out there, Parker, and take some pictures. That's what I pay you for."

"Pictures of brats playing? The guy's gotta be crazy", muttered Jameson as Peter left the office.

Peter was fed up. And his day was not about to improve. His exasperating meeting with Jameson had made him late for class – again. Peter hot-footed it to campus and ran towards his class. But he was too late. The hapless student ran headlong into his teacher, Dr. Connors.

"Parker," roared Dr. Connors, "watch where you are going. What is it with you, young man? Class is over. And now that I do have your attention, where is your paper? It is late and if it doesn't come in soon, you will fail. I don't understand you, Parker. You used to be such a diligent student, but these days

your mind does not seem to be on your studies."
Peter could hardly explain that he spent much of his time protecting the city from creepy crooks and venomous villains. How was he ever going to manage his life as Spider-Man, super web slinger, with the life of studious student? Peter hung his head and stammered out an apology.
He promised Dr. Connors that he would get his science paper in on time.
"Oh no!" Peter glanced down at his watch. He was going to be late for Aunt May. For the second time that day, Peter ran as fast as he could. Flustered, Peter rushed into Aunt May's house.
"Surprise!" Peter looked around to see his close friends and aunt looking excited and expectant.
"What's the surprise for?" Peter was confused. Aunt May looked at him in disbelief. "Peter, it's your birthday. Surely you can remember that?"
"Peter lives in another reality, don't you Pete? Long time no see," quipped Mary Jane. Peter could hardly ignore Mary's sarcastic tone, and it hurt him deeply.
For a long time, Peter had loved Mary Jane deeply. But apart from one kiss, their friendship had never gone beyond that of good friends. Peter knew that a close relationship with him could

put the girl he loved in danger. Once before, a monstrous madman called the Green Goblin had threatened Mary Jane. Peter was determined that his double life as Spider-Man should never again put his sweetheart in danger.

"Well, he's probably too busy taking pictures of that pestilential Spider-Man," said Harry. Peter knew that his friend Harry loathed and detested Spider-Man. Harry blamed the webbed wonder for the terrible, agonizing death of his father, Norman Osborn. But only Peter knew that Norman Osborn had in fact been that same gruesome Green Goblin who had threatened Mary Jane and terrorized the city.

The mood at Peter's party was definitely chilly. This was not turning out to be the best birthday in the world.

"Any more cake, anyone?" asked Aunt May cheerily. She wanted to change the subject. "Let's have a toast to all the wonderful things that are happening to us. Such as Mary Jane being in a new Broadway play."

Mary Jane looked a little embarrassed. "Actually it's an 'off-Broadway' play, Aunt May", she corrected.

Determined to have a jolly party, Aunt May continued to

celebrate their achievements. "And Harry, your father would have been so proud of all your wonderful successes at OsCorp. I only wish he was here to see it." OsCorp was the company run by Harry's father, where Harry became Head of Special Projects after his father's death.

"Peter, you've heard of the scientist, Dr. Otto Octavius, haven't you?" asked Harry.

"Dr Octavius?" spluttered Peter in between mouthfuls of birthday cake. "He's my hero. In fact, I'm writing my paper on the great man."

"Well, OsCorp is funding Dr. Octavius' latest research. He's due to give a demonstration soon that he promises will be pretty spectacular. I'll introduce you to him, if you like," offered Harry. "Come up to OsCorp and I'll fix it up."

At last. Peter's birthday was looking up. "Fantastic," he exclaimed.

Mary Jane looked at her watch. "I must be going Aunt May", she said. "I'm sorry to break up the party but I have to get back to the city for the performance tonight."

Harry offered to give Mary Jane a ride in his car. Peter then promised his favourite girl that he would be at the show tonight. Mary Jane looked sceptical. "I'll believe it when I see it," she said dismissively.

After Mary Jane and Harry had left, Peter sat down with his Aunt May. He had some ice cream and birthday cake. Aunt May and Peter chatted about what had been happening. It was some time since they had spent any time together. Aunt May lived alone. Her husband, Ben, had been killed by an unscrupulous carjacker – the very villain that Peter had let go. Since Ben's death, Aunt May was struggling to keep on top of mounting debts. Shortly, Peter told Aunt May that he should be on his way, to make sure he wouldn't be late for Mary Jane's show.

Peter wanted to surprise Mary Jane with some flowers. He stopped off at a shop on the way to choose a bunch. Just as he was about to hand over his money, Peter's spidey sense alerted him to danger. Peter heard gunshots. He turned around and saw a car speeding away. No time for flowers. This was Spidey time.

Peter dived into a dark deserted alley where he speedily changed into his tight-fitting, eye-grabbing Spider-Man suit.

PING, PANG, BANG! Suddenly Spider-Man was under siege from a hail of bullets. With the speed of lightning, the wonder bug dodged the deadly missiles. As the web slinger bounded onto the villains' car a sticky mesh web, strong as steel, light as wool, zapped out of his hand. With a quick spin, the criminals were all wrapped up.

"Wanna go for a spin?" he jeered. The hapless villains found themselves being bounced out of their get-away car by Spider-Man's sticky web. Stunned by what was happening, the wrong-doers were powerless.

Spider-Man caught crooks like a spider catches flies. He left the villains hanging unceremoniously on a lamp post, waiting for the police to deal with them. "Hang loose," shouted Spider-Man, fleeing before the police arrived.

"Oh no," Peter realized with horror that he was late for Mary Jane's play. "And I promised her I'd be there."

While Peter had been diving from bullets and dangling with baddies, Mary Jane had been watching his empty seat in the theatre. With increasing irritation and disappointment, Mary Jane wondered where Peter could be, or even if he was coming.

For the fourth time that day, Peter was late. He arrived at the theatre in a beat up convertible. His clothes were disheveled and he looked a mess. But he was too late. The theatre had closed its doors. The play had begun. Peter begged the usher to let him in, but the usher was insistent. No latecomers were to be admitted.

Peter was dismayed. He had let Mary Jane down, again. How could he carry on like this? There was no way Peter was going to be able to live as a normal guy and as crime-buster Spider-Man. Peter longed to tell Mary Jane about his secret life. But how could he?

Peter returned home. He couldn't sleep. He kept thinking about Aunt May and Mary Jane. Every time he closed his eyes, he heard Uncle Ben's words ringing in his ears: "With great power, comes great responsibility."

That night, Peter rang Mary Jane. Even though Mary Jane was at home, she didn't pick up the phone. She listened with scorn as Peter stammered out an apology for missing the play.

But Mary Jane had heard too many excuses from Peter Parker. She had had enough. She turned off the answer machine while Peter was still talking. Mary Jane never heard Peter as he mumbled out his admission that he was, in fact, Spider-Man.



*story continues on page 29*

# Crack the Code

**Peter** has a date with **Mary Jane.** But unfortunately he's called into Spidey action to deal with **Doc Ock**.

The webbed wonder has managed to send a coded message to **Mary Jane.** Can you help her crack the code? You'll need a pencil and paper.

23 1 9 20 6 15 18
13 5. 9'13 9 14 1
2 9 20 15 6 1
20 1 14 7 12 5.
16 5 20 5 18 24 24 24

**Use the code cracker below to help:**
**A1 B2 C3 D4 E5 F6 G7 H8 I9 J10 K11 L12 M13 N14 O15 P16 Q17 R18 S19 T20 U21 V22 W23 X24 Y25 Z26**



Answer on page 60.

# Imposter Alert!

**Doc Ock** has created a villainous **Spider-Man** clone. Look closely at the pictures and descriptions of both **Spider-Men**. Can you work out who is the real hero and who is the morphed impostor? Their are 7 differences between them.Check out the clues if you need help.

**The Real Spidey:**
- traps his enemies in webs
- can climb walls
- believes responsibility comes with great power
- looks after everyone

**The Imposter Spidey...**
- uses claws to attack his enemies
- flies like a bird
- believes wealth comes with great power
- saves only those who can pay



Turn to page 60 to find out if you are right.

## Chapter 2

Everything was in place. After years of research and experimentation, Dr. Octavius was ready to demonstrate to his peers and critics his biggest achievement to date. It was the day the scientific community would be stunned by his demonstration. Among many prestigious and important scientists gathered to witness Dr. Octavius's promise of spectacular scientific advancement, were Harry Osborn and college friend and science enthusiast, Peter Parker.

Dr. Octavius welcomed the guests to his laboratory. When they were all assembled, the brilliant scientist opened the drape. Behind it lay an extraordinary machine with octopus-like arms connected to a harness. To the amazement of the guests, Dr. Octavius stepped into the machine. As he did so he explained to the curious crowd that the octopus arms were 'smart arms'. They were designed and built to help humans in multi-tasking or working in dangerous conditions. The possibilities were endless.

"Just think, across the globe, surgeons could perform a surgery alone. Workers would never have to personally handle radioactive materials or other dangerous substances. The uses are boundless!" enthused the great man.

With mounting confidence, Dr. Octavius showed off the incredible variety of tasks the arms could complete. The invited guests were suitably amazed. Encouraged by their gasps of wonder, the doctor demonstrated more and more tasks that these super robotic tentacles could complete.

But something was not right. The arms were not obeying the instructions. The demonstration was going horribly wrong. The arms were malfunctioning. They swung wildly out of control, knocking over anything and anybody in their way. The guests' amazement soon turned to terrified panic.

Dr. Octavius was determined to rescue his reputation. He shouted that all was under control. But clearly it was not. The arms ripped out parts of the lab. CRASH SMASH!

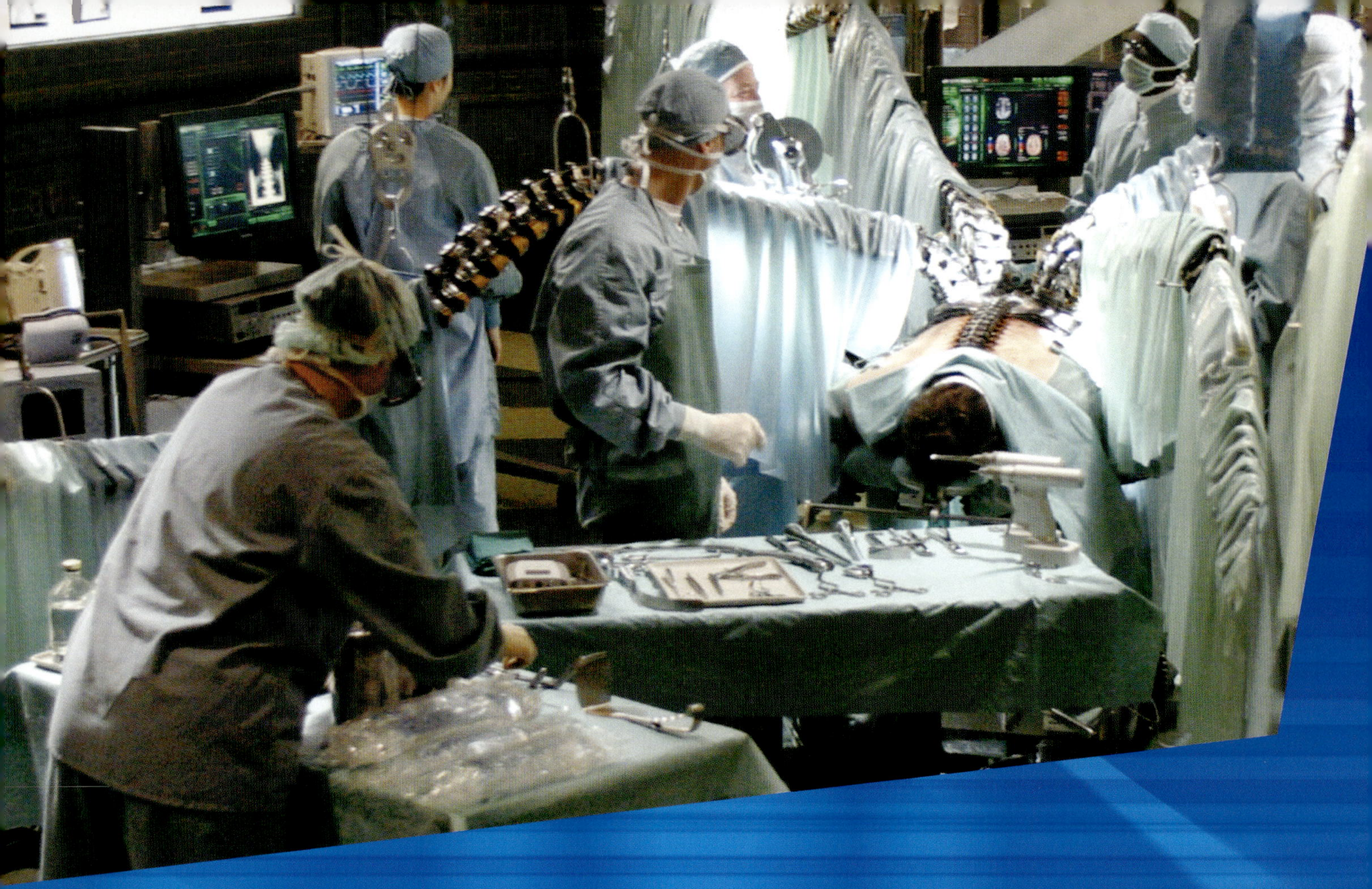

Furniture and equipment toppled and smashed. Guests hid wherever they could to avoid being crushed by a long strong maniacal robotic arm. The doctor had no control over his invention. These tentacles had taken on a life of their own.

No-one noticed Peter quickly leave the room.

Spider-Man swung into the lab. Harry, who had long harboured an intense hatred of the creature that he believed caused his father's death, shouted out that the disastrous experiment was the fault of Spider-Man. Despite such unfair accusations, Spider-Man swept up Harry in his web, just in time to save Harry from a flailing tentacle.

The tentacles were smashing into everything and everyone. One abominable arm crashed into the electrical power supply. Sparks and flares shot out everywhere. A powerful electrical current ran through the tentacles and straight into the powerless doctor's back. In a shower of lightning flashes and explosions, the tentacles fused onto the scientist's spine. In agony, Dr. Octavius writhed and thrashed around. The massive electrical surge relentlessly pulsed through his body. Spider-Man realized what a terrible thing was happening to his scientific mentor. He leapt over the power source and turned it off. The flashes petered out. The electricity supply ceased to wrack the doctor's body. As the poor man fell to the ground, he saw Spider-Man standing by the electricity supply.

It seemed as if Dr. Octavius was dead. The room was hushed. Then suddenly, one of the tentacles reached up into the air, as if coming to life.

In the dead of night the most experienced surgeons stood over

Dr. Octavius' body, examining the extraordinary arms. The medical team had decided to try and cut off the mechanical arms so they could save the scientist's spine. A surgeon turned on the saw. Without warning, the tentacles reared up and attacked the surgical team. Doctors and nurses were hurled across the surgery room by these monstrous mechanical arms.

The massively heavy pull of the tentacles wrenched Doc Octavius to a sitting position. As he looked around him at the chaos in the hospital room, flashes of what had occurred earlier that day came to him. The demonstration. The smart arms malfunctioning. People running for cover. The lab being trashed. And Spider-Man hovering over him with his hand by the electrical supply.

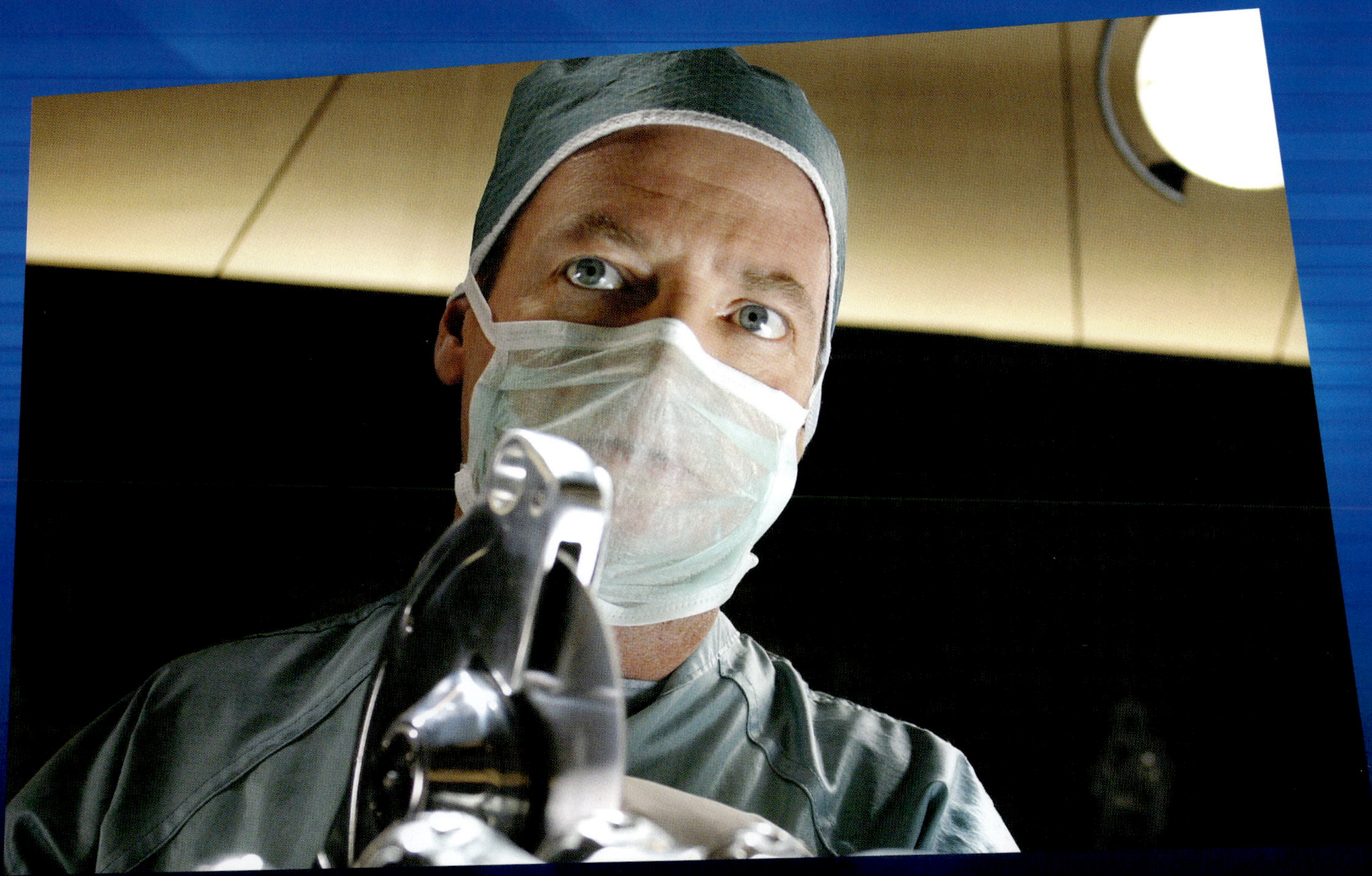

ARRRGH Screaming in agony and despair, the man with maniacal arms fused to his poor mangled body managed to pull himself out of the bed. With the tentacles stuck to his body like blood-sucking barnacles, the scientist fled from the hospital.

In a stupefied daze, Dr. Octavius lurched through the dark night to his lab. He looked in horror at the devastation. His life's work was ruined.

Blaring sirens jolted the doctor out of his dazed horror. He suddenly realized that he was a wanted man. The police were

after him. This was no place to stay. Where could he flee? Looking out of a smashed window, the doctor spotted a deserted and dilapidated pier. Could he find refuge there for a while?

The desperate man escaped to the sinking pier. He stood in the cover of darkness, thinking about the terrible events that had shattered his life. But his thoughts changed. This was no longer Dr. Otto Octavius. He was no longer himself. The tentacles were taking over his brain and nervous system. They were telling him that the catastrophic events were due to one man only – Spider-Man. That wretched web-weaver had ruined his dream. But the scientist would not give up. No, he would be given the recognition he deserved. He would rebuild his lab.

The experiment had gone perfectly. The arms were perfect. But he would need money for his new plan. Triumphantly, the deluded scientist raised his tentacles into the air.

He was Doc Ock.



story continues on page 35

# All in a Twist

**Mary Jane is in a tangle,** and not with Spidey's web.

**Doc Ock** has got her by his trailing tentacles. **Spider-Man** is desperate to save his favourite girl-next-door but can't figure out how to reach her.

Can you **help Spider-Man rescue Mary Jane** from the terrible twist?

Answer: page 61

# Webbed Up

A spider has been busy catching lunch. How many flies can you see hidden in this intricate web?



Answers: page 61

## Chapter 3

Peter was out with Aunt May, helping her with her weekly shop at the grocery store. Peter was carrying two rather awkward and heavy shopping baskets, trying to find Aunt May who had disappeared down a different aisle, looking for flour so she could bake some cakes later that day.

Uh oh. Peter's skin began to tingle. Something was up.

He heard Aunt May screaming as she struggled to hold onto her purse. "Get off me, you brute" the elderly lady yelled. Peter dropped the shopping baskets and disappeared.

Spider-Man leapt into view. Customers looked astounded as he weaved his way through the aisles to save Aunt May from a dastardly robber.

He found Aunt May standing triumphantly over the mugger, her foot pressing hard into his stomach. "Arrrgh. Get her off me," moaned the mugger. Aunt May had thumped the mugger so hard with her bag that he had fallen to the ground. "Nice work, m'aam," said Spider-Man as he pulled up the mugger and hauled him off to the security guard. Before Aunt May had a chance to tell Spider-Man how she had managed to foil the villain, Spider-Man was off.

"Peter, where did you get to?" said Aunt May crossly. "Didn't you see that awful mugger try to take my bag? And why did you drop my shopping everywhere?" Peter said nothing, but helped pick up the shopping. He listened as Aunt May told him how she had knocked out the mugger. "And that fine Spider-Man told me I had done brilliantly", she said proudly. "Pity he didn't hang around for a chat," she continued wistfully.

Peter gave a wry smile.

Peter was walking in the city. His journey took him nearby to a bank. Spider sense made Peter look inside. A man was in a long trench coat. Not an unusual sight. But with four monstrous long violent tentacles snaking their ugly way from under the coat? This was no ordinary man. The tentacled-beast exploded through the bank vault.

In an instant, bells were ringing. Alarms flashed. Guards came running from everywhere. Terrified customers dived for cover. A courageous bank guard confronted the beast. But to no avail. With the twitch of a tentacle, Ock tossed the unfortunate man into the air, as if he was a rag doll.

Just as the tentacled terror was reaching into the vault and grabbing bags of money, Spider-Man careered into action. Leaping towards the villain, he shot webs, crawled walls and jumped into the air, trying to avoid Ock's crushing arms. Suddenly, the tentacles grabbed Spider-Man around the throat.

Struggling to breathe, he gasped for air.

Using all his power, Spider-Man fired a web onto a massive desk across the room. He pulled the web towards him and slammed the heavy desk into Ock's tortured body. The table smashed into Ock with such force that he was thrown out of the window. With lightning reflexes, Spidey followed him into the street.

A terrible fierce battle followed. Spider-Man used all his strength and power but the force of Ock was too much. Spider-Man had been defeated.

As Ock triumphantly left the scene of the battle, Spider-Man defiantly shouted, "You can run doc, but you can't hide."

Peter made his way back to his apartment. Still wearing his Spider-Man suit, he looked terrible. His face was bruised and bloodied. His body ached with every move. He had taken a hard bashing from Ock's mad tentacles.

Peter looked at the framed photographs of his beloved ones. Uncle Ben, Aunt May and Mary Jane. In another corner of the room, he saw a pile of school books on his desk. None had been opened by the student. Nearby, there were stacks of unpaid overdue bills. Peter's life was a mess. He was no longer a worthy friend or loyal nephew. He wasn't even a good student any more.

Tired and forelorn, Peter looked down at his torn Spider-Man suit. He could bear it no longer. Ripping off the suit he shouted, "This isn't worth it." Peter Parker stuffed the red and black suit in a box. He pushed the box as far back into the cupboard as possible.

The light on the answer machine was blinking. Furious and frustrated at the hand bite of fate, Peter tapped the play button. It was Aunt May.

"Hi there Peter. It's Aunt May. Would you be able to help me move some furniture tomorrow. I can move some of it, but some is just too heavy. Hope your studies are going well. Hear from you soon."

The next day, Peter rang Mary Jane to ask if she would meet him for a coffee. Mary Jane agreed to meet her elusive friend.

But first, Peter had arranged to meet his old friend Harry. Peter didn't want to talk about what had happened at Dr Octavius' lab. But Harry would talk about nothing else. "That wretched Spider-Man. It's all his fault. He ruins everything" moaned Harry.

"But, but, didn't he scoop you out of danger?" protested Peter.

Harry ignored that remark but continued to vent his anger about the Spider-Man.

"Spider-Man is bad news. He killed my father and I'll get my revenge somehow," he vowed.

"But Harry", said Peter, "your father was...,"

"My father was what?",snapped Harry. "Why are you always so quick to defend Spider-Man, Peter? What do you know about him?" quizzed Harry. Peter knew he had already said too much. "It doesn't matter," he said.

Harry looked angry. Peter left with a heavy heart. Harry was eaten up with hatred for Spider-Man.

Mary Jane and Peter met in a café. Very soon, Peter poured

his heart out. "Things are going to change Mary Jane. I feel it. I realize now how awful things have been. I'm so sorry. I've let you down, and you are so important to me."

Mary Jane listened as Peter said he would be a better friend, and would always be there for her, as he had once promised. Mary Jane smiled gently. She was so glad that Peter had decided to turn over a new leaf. She had missed the Peter Parker she knew and loved.

"But, as part of the new me," Peter said as he pushed his chair back, "I have to help Aunt May move some furniture and I don't want to be late. I'll talk to you soon, Mary Jane."

"I can't wait," she smiled.

Bang on time, Peter arrived at his Aunt May's. It felt good to have discarded his alter ego, super-hero status. Now Peter

could concentrate on getting his own life back together.

Peter and Aunt May moved the furniture and chatted.

"Oh, Peter", said Aunt May, "do you have any pictures of that terrible fight Spider-Man had with a horrible man with these monstrous arms, Doc Ock, down at the bank?"

Peter stopped in his tracks. How on earth did Aunt May know about the battle at the bank? Trying hard not to show his surprise, Peter asked Aunt May how come she knew about Spidey and the tentacled terror.

"Oh, well," began Aunt May, "I have two girlfriends who work down at the bank. And do you know, Peter, that Spider-Man saved their lives. If it wasn't for that man, those dear friends would not be here today. I wonder if Spider-Man knows just how much the people in this city think of him. But we never get a chance to tell him how grateful we are."

Peter smiled. He had forgotten how Spider-Man appeared to the good folk of the city. They liked him. They needed him! With a blinding flash of clarity, Peter realized that he could do it. He could be plain old Peter Parker and he could use his super powers wisely. Peter no longer wanted to discard the hand that fate had dealt him. Indeed, he was determined he could be an unflinching loyal friend and an intrepid hero. He felt great.

Spider-Man was back, and here to stay.

***story continues on page 44***

# Double

**Arrgh – it can't be!**

There are two devilish Dr Ocks.
He has cunningly morphed a copy of himself.

But there are eight differences between the real tentacled terror and the copy.

**Can you find them?**

# Trouble

Answers: page 61

## Chapter 4

The door to Harry Osborn's apartment smashed to the ground. To Harry's astonishment, Ock strode into the room, tentacles waving madly and wildly in the air. Harry took cover wherever he could.

"Give me money, Osborn," demanded the maniacal madman. "I need money to complete my research."

Harry was furious at the rude invasion of his privacy. He was also amazed that Ock could even consider that he would get money to finish his mad-capped research. "You've got to be kidding" said the young man. "No way am I going to give you money." But even as Harry was speaking, a plan was formulating in his mind. What if he paid Ock to kill his arch enemy, Spider-Man. This was the man responsible for his father's death. That would be money well spent.

Ock raised one of his massive tentacles. But before it could crash down upon Harry, the young man stood and shouted out.

"Unless," continued Harry, "you kill... Spider-Man. Do that, and you can have all the money you need for your wretched research."

Ock agreed instantly:"Deal. Just tell me how I find him?"

"That's easy. Follow Peter Parker," said Harry. "He'll lead you to him."

That night, Peter drove his scooter back into the city. He'd had a good day with Aunt May. But Peter felt he wasn't alone. Someone was watching him as he turned corners or stopped at lights. He was never far away from prying eyes.

Without warning, a tentacle swooped across the city sky and plucked Peter from his scooter like an octopus scooping up a tiny fish to devour. Hurtling through the air on the tentacle, Peter was brought face to face with the menace to mankind, Doc Ock.

The madman growled at the young photographer. "You tell Spider-Man to meet me at Pier 56 or I'll hunt you down and kill you." Dangling on a tentacle, Ock dropped Peter to the ground. Landing with a bump, Peter watched Ock flee, tentacles twitching behind him.

There was no time to lose.

Spider-Man swung into action. He leapt across to the pier. Ock was fired up by a determination to kill the wretched hero. With the death of the webbed wonder, Ock could lay his hands on all the money he needed. But Spider-Man was not to be beaten a second time. Spider-Man used all his resources. He dodged, jumped and leapt. He fought for his life.

Pieces of metal went flying. Planks of wood shot through the air. The quiet dead of night was pierced by terrible screams and thuds as the two battled against each other.

After a long and terrible battle, Ock lay crumpled in a heap. His tentacles no longer taunted and twitched. They were lifeless clumps of metal. The pier was a wreck. Ock was defeated. Spidey had saved the day.

Meanwhile Mary Jane was drumming her fingers, looking at her coffee, wondering where her friend was. She had arranged to meet Peter but... Rushing in, Peter arrived. He was late, but at least he had made it.

Peter looked at Mary Jane. "I need to explain," he started. "I said things were back to normal, but I was wrong. Life isn't simple for me. Mary Jane, I won't be around for you as much as I want."

"But you do want to be around", asked Mary Jane. Peter looked at her lovingly. "Of course I do," he said. The two held hands across the table. Peter and Mary Jane agreed to work harder at keeping their special friendship.

Life for Peter Parker was looking up.

And Spider-Man was here to stay.

***The End***

# Web Word

Can you work out the answer to each spidery clue? Fill in the grid below.

**ACROSS**

**1** A spider belongs to this group of animals (8 letters)

**5** Dr. Octavius conducts this as part of his scientific research (10 letters)

**8** Mary Jane has always wanted to be one (7 letters)

**10** What a spider weaves (3 letters)

**11** The first name of Peter's best friend (5 letters)

**DOWN**

**2** These are fused onto Dr. Octavius' body (9 letters)

**3** What Spider-Man fights against (5 letters)

**4** The name of Peter Parker's uncle (3 letters)

**6** Peter is late for his own one (5 letters)

**7** The name of Peter Parker's aunt (3 letters)

**9** The doctors try using one of these to cut off Ock's tentacles (3 letters)

Answers on page 60.

# Face Off!

Get under the skin of the web crawler or try being a tentacled terror with these easy-to-make face masks. Make a mask for a friend and see how you pit your wits and strength against each other.

You'll need a piece of card for each mask, tracing paper, pencil, scissors and string.

Trace over a face below. Turn the tracing paper over and pencil over the line again. Now turn the tracing paper back again and pencil over the drawing onto the card. Cut out the card. Paint the Spidey mask in black and red colours. With the scissors, pierce a hole in either side of the card. Thread a piece of string (long enough to go round your head) and ask a friend to tie the mask on you. Away you go!

# Spider Spin

**Did you Know?**
Spiders are not insects. True insects have three main body parts – head, thorax and abdomen.

Spiders belong to a group of animals called arachnids. They have two main body parts made up of a combined head and thorax, and an abdomen. Other members of the arachnid group are ticks, mites and scorpions.

**Colour of Blood!**
If Spider-Man cuts himself he bleeds – with red blood. A real spider's blood is pale blue!

**Grab a Bite!**
Spider-Man uses his web to catch creepy crooks. Real spiders can use their webs to catch an insect or two for dinner!

**Ballooning is A Breeze**
A breeze can carry a young or small spider through the air as it hangs onto a long strand of silk from one of the spinnerets on its abdomen. This is known as ballooning.

**Don't Scream!**
If you see a spider in the bath, try not to scream! Most spiders don't bite humans unless they are frightened.

**Ambush**
Different types of spider have different ways of catching food. Some lie in wait, on plants, tree bark, on the ground or under stones.

When they spot a passing grub – they pounce! Other spiders don't hang around waiting – they simply go hunting for prey.

**Liquid Lunch**
A spider cannot eat solid food. Instead, it injects venom into the victim, paralyzing it and turning its organs to liquid. Then the spider sucks up the liquid from the victim... Slurp!

**Silky Strong**
Spider webs are made of long strands of spider silk.

These are produced from about six silk glands called spinnerets beneath the spider's abdomen. Liquid spider silk squeezes out of spinnerets. The liquid thread hardens as it leaves the spinneret.

**True or False**
There is a goat that is part spider?

**True!**
Scientists have combined the DNA from a goat and spider to create a goat that is 1/70,000th spider.

With the spider silk protein gene in its body, the goat produces a fibre from its milk that for its size is five times stronger than steel! The fibre called BioSteel® and is light but incredibly strong.

**Roll Away**
If Spider-Man ever found himself in a spot of bother in the hot desert, he could copy a wheel spider. If anyone or anything attacks this arachnid, it simply tucks its legs into its body and rolls away.

**Most Unwanted Spiders**
You don't want to be bitten by one of these:

- Black Widow – (US)
- Wolf spiders (South America)
- Running spiders (worldwide)
- Red Back spider (Australia)
- Funnel spider (Australia)
- Recluse spiders (US)

**Jump!**
Jumping spiders can leap up to forty times their body length.

**Spitting Spider**
Spitting spiders immobilise prey by spitting at it. **Uck!**

**Shy but Mean**
The black widow spider is shy, usually staying in her hidden web. But watch out – if her web is disturbed, or she becomes trapped in a shoe or piece of clothing, she may bite, injecting the hapless victim with a paralyzing poison. The adult male is harmless. After mating, the female usually eats him!

**It's a Trap**
A trap-door spider is very crafty. It digs a burrow and covers it with silk and earth. When a tasty insect wanders by, the spider rushes out of the burrow, poisons the unfortunate creature and drags it into its hole to eat.

**Spidey Sense**
Even though most spiders have eight eyes, they don't have brilliant eyesight. But super sensitive hairs on a spider's body alert the spider to any movement around them.

# Make a Copy

**Aunt May wants to show Spider-Man that the city appreciates his crime-busting efforts.**

She is making posters of Spider-Man to put up all over the city. Can you help?

Use the grid below to copy the picture of **Spider-Man**. Don't forget to colour it in.

Thanks Spidey!

# THE CHASE IS ON

Pit Dr Ock and Spider-Man against each other in this game of skill and chance. Who will be the first to reach safety at the end of the pier, and who will be left lingering and out for revenge?

**How to play**
You'll need a die, and two counters and a friend to play with.

Decide who will play our favourite webbed wonder and who will be the fantastically foul Dr Ock. Roll the die to decide who starts the game. Then take it in turns to roll the die and follow the instructions on the board. If Spider-Man lands on a purple square, he must miss a go. If Ock lands on a red square, he must miss a go.

Now, webbed wonder and manic menace, play on. **May the best, or worst, win!**

START
Don't forget Ock, if you land on red, miss a go. Spidey, you carry on!
1
Spidey, don't forget if you land on purple, miss a go. Ock, you carry on!
2
Well done, Spidey. You rescued two children from danger. Move forward 2 spaces.
3
Bad luck Ock. One of your tentacles is caught in a door. Go back 2 spaces.
4
5
Need a helping hand, Ock? Move forward 2 spaces.
6
7
8
9
Watch out Ock! A crazy police dog is patrolling the bank. Go back 2 spaces.
10
Spidey, you've ripped your clothes. Go back 4 spaces
11
12
Spidey, go back 6 spaces. This is a dead end.
13
14
Good deeds deserve rewards Spider-Man. Move forward 6 spaces.
15
16
17
Ock, raiding a bank is no way to behave. Go back 4 spaces.
18
Spidey, can you hear MJ calling? Check it out. Move forward 2 spaces
19
Ock, it's your lucky day. The coast is clear. Move forward 4 spaces.
20
21
22
Darn it, Spidey, you forgot to visit Aunt May. Go back 6 spaces.
23
24
25
So nearly the winner. But you've fallen into a trap laid by Spider-Man. Go back 4 spaces.
26
Watch that loose plank, Spider-Man. Go back 2 spaces.
27
FINISH!
You're the first to reach the end of the pier. But watch out – the enemy is catching up!

# Ultimate Spider-Fan

**1.** What is Dr Octavius' latest research into?

**2.** How many tentacles fuse onto Ock's body?

**3.** MJ stands for Mary Joan. True or false?

**4.** Harry blames MJ for the death of his father. True or false?

**5.** Who told Peter that "with great power comes great responsibility"?

**6.** Spider-Man's spidey sense warns him of danger. In actual spiders, which body part helps warn the spider of danger?

**7.** What does MJ want to be?

**8.** What is Peter's job at the Daily Bugle?

**9.** J Jonah Jameson is MJ's father. True or false?

**10.** Spider-Man has five fingers on each hand. True or false?



Turn to page 60 to see how well you know your Spider-Man.

# Joining Up

You'll need a pen and patience for this. There are over 50 dots to join together before you can work out what is up! Colour it in when you're dotted out.

# Activity solutions:

Page No.14
Activity: Caught in a web
Solution:

| | | | | | | | | | | |
|---|---|---|---|---|---|---|---|---|---|---|
| B | D | U | L | L | A | W | U | E | C | D |
| C | M | T | A | V | P | G | P | I | R | N |
| L | T | F | J | X | D | K | C | W | A | L |
| H | U | R | M | O | A | Y | L | E | W | I |
| R | E | S | P | O | N | S | I | B | L | E |
| S | N | R | T | U | G | A | M | L | E | G |
| R | I | D | O | N | E | O | B | B | R | C |
| Q | U | Y | L | C | R | J | E | V | S | V |
| B | P | T | P | O | W | E | R | E | U | Z |
| U | W | F | K | R | E | N | N | I | P | S |

Words:
Web
Spinner
Power
Responsible
Wall
Climber
Danger
Hero
Crawler

Page No.15
Activity: Make a match
Solution: A&L, B&G, C&I, D&F, E&J, K&H

Page No.27
Activity: Crack the code
Solution: Wait for me. I'm in a bit of a tangle peter xxx

Page No.28
Activity: Imposter alert
Solution: Did you spot the 7 differences?
Blue boots not red
Steel claws
Extra strip around waist
Green eyes
7 legged spider logo on chest
Blue neck
Distorted web pattern on mask

Page No.33
Activity: All in a twist
Solution: Tentacle E leads to MJ

Page No.34
Activity: Webbed up
Solution: 10 flies in the web

Page No.42
Activity: Double trouble
Solution:
1. Dr Ock has lost an ear
2. Top right tentacle has a blue light
3. Dr Ock has grown a extra tentacle
4. Bottom right tentacle claw has moved
5. Dr Ocks coat is wider on the right
6. Top left tentacle has a fourth claw
7. The largest claw has an end missing
8. The lapel on the jacket is different

Page No.49
Activity: Web word
Solution:

ACROSS
1 Arachnid
5 Experiment
8 Actress
10 Web
11 Harry

DOWN
2 Tentacles
3 crime
4 Ben
6 Party
7 May
9 Saw

Page No.58
Activity: Ultimate Spider fan
Solution:

1) Dr Octavius' research is into fusion as a source of energy
2) 4 tentacles are fused onto Ock's body.
3) False. MJ stands for Mary Jane.
4) False. Harry blames Spider-Man for the death of his father.
5) Peter's Uncle Ben told him "with great power comes great responsibility"
6) The hairs on a spider's body can detect if someone or something is nearby and alert the spider to danger.
7) MJ wants to be an actress.
8) Peter is a photographer for the Daily Bugle.
9) False. J Jonah Jameson is the editor at the Daily Bugle.
10) True.

Page No.59
Activity: Joining Up
Solution: